A field guide to the geomophology of the Slapton region

by Tim Burt

School of Geography Oxford University

Contents

The climate and hydrology of the Slapton region. *Tim Burt*	3
The geology and structure of the Slapton region. *Keith Chell*	5
Weathering and soils. *Trudgill*	9
Slope hydrology and geomorphology. *Tim Burt*	12
Sediment and solute dynamics. *Burt and Louise Heathwaite*	15
Sediment yields and budgets in the Start valley. *Ian Foster, King and Phil Owens*	25
Lake sedimentation. *Heathwaite*	31
Landforms and Quaternary history of the Prawle coast. *Mottershead*	42
The Start Bay barrier beach *David Job*	47

Chapter 1
The climate and hydrology of the Slapton region

Tim Burt

School of Geography
Oxford University

Slapton lies within that part of Devon to the south of Dartmoor known as the South Hams. Any study of landforms within this area must immediately take two points into account. First, the highly folded Devonian strata (Chell, this volume) have clearly been subject to planation, probably during the late Tertiary. This episode necessarily provides the starting point for any subsequent denudation chronology. Second, this part of England was never glaciated, though its proximity to the Pleistocene ice sheets to the north must have given it a truly periglacial climate. As Trudgill and Mottershead (this volume) both point out, deep mechanical weathering of local bedrock and the impact of solifluction processes have been dominant features of the Pleistocene in this part of England. A detailed consideration of the morphology of the Slapton region is provided by Mercer (1966).

A Meteorological Office climatological station was established at Slapton Ley Field Centre (32 m O.D.) in the spring of 1960; observations are still made daily. The records up to the end of 1973 were the subject of a paper by Ratsey (1975). Table 1.1 lists climatic data for the Centre. The current climate is remarkably mild: an overall mean temperature of 10.5°C with a January mean of 5.8°C and a July mean of 15.8°C. Really cold days, below 5°C, are rare. Slapton boasts an almost frost-free climate with only 22 air frosts (range 8-43) and 56 ground frosts on average. Ratsey notes an absolute minimum of only -5.6°C (12.1.63; 8.2.69). The growing season (temperatures above 6°C) lasts from early March to Christmas. Plentiful rainfall occurs throughout the year although, as with any British location, there is a good deal of variation month to month. Mean annual rainfall (1961-90) is 1034 mm; rainfall is lowest and most variable in

Table 1.1 Climatological and hydrological data for the Slapton region.

	mean temperature (°C)	mean maximum temperature (°C)	mean minimum temperature (°C)	mean rainfall (mm)	mean runoff (mm)
January	5.8	8.2	3.3	124	99
February	5.7	8.1	3.3	101	115
March	6.8	9.8	3.6	92	58
April	8.9	12.5	5.8	62	37
May	11.3	14.6	7.9	65	18
June	14.3	17.7	10.7	59	10
July	15.8	19.7	11.9	58	8
August	15.7	19.2	11.9	72	7
September	14.6	17.8	11.2	79	12
October	12.4	15.3	9.6	92	19
November	8.5	11.3	5.7	108	42
December	6.7	9.3	3.8	122	58
Year	10.5	12.8	7.4	1034	484

Temperature data 1960-73, from Ratsey (1975). Rainfall data, 1961-90. Runoff data, 1971-2, 1974-78, 1983-5.

spring and early summer, and highest in December and January. Since 1960 there have been 11 months with over 200 mm and only 11 months with less than 20 mm, four of them in 1976. The wettest day on record is 23 July 1987 when 88 mm fell in 5 hours. Snow is rare, usually less than three occasions per year, though about once a decade Slapton, like the rest of south Devon, may be affected by severe snowfall. Often this happens when most of the UK is subject to the influence of the Scandinavian anticyclone; fronts skirting its southern limb can produce large amounts of snow in south west England. Given this mild, moist climate, one concludes that weathering activity must be dominated today by chemical processes (see Trudgill, Chapter 3, this volume) and that frost action is now rare in this area.

Despite the relatively uniform rainfall, runoff is strongly seasonal (Table 1.1). This is in part caused by the strong evaporation losses of the summer months but also because the runoff is largely subsurface in origin. Indeed, the fractured, upper part of the shale bedrock (itself an aquiclude) functions like a shallow aquifer. Soils dry out quickly in the spring and percolation does not recommence in autumn until soils have returned to field capacity. Streamflow is very low in June, July and August (less than one quarter of the mean flow) showing how restricted is the groundwater supply from the fractured shale; even so, most streams are perennial. In contrast, in January and February flows are at least 2.5 times the mean flow (Burt, 1992). Blackie (pers. comm.) provided the following water balance for the Slapton Wood catchment for the 1990 water year:

Precipitation	1015 mm
Runoff	540 mm
Actual evaporation	475 mm
Potential evaporation	767 mm

Blackie showed that rainfall for the Field Centre was 88% of that for the Slapton Wood catchment. Van Vlymen (1980) argues that rainfall for the Slapton Ley catchment as a whole is underestimated by 15-20% by the Centre gauge. He gives the following runoff figures for the Slapton catchments (Figure 5.1):

Gara	781 mm
Start	535 mm
Slapton Wood	463 mm
Stokeley Barton	294 mm

An increase in rainfall and a decrease in evaporation towards the higher, northern part of the basin account for this variation, though the low runoff from the Stokeley Barton stream may be caused in part by deep seepage.

The hydrology of the Slapton catchments has been considered in detail by Troake and Walling (1973) and by Van Vlymen (1980). Further information on hydrological processes is given in Chapter 4.

REFERENCES

BURT, T.P. (1992). *The hydrology of headwater catchments*. In: P. CALOW and G.E. PETTS (eds), The Rivers Handbook, Volume 1, Blackwell, 3-28.

MERCER, I.D. (1966). The natural history of the Slapton Ley Nature Reserve I: Introduction and morphological description. *Field Studies* 2, 385-405.

RATSEY, S. (1975). The climate of Slapton Ley. *Field Studies* 4, 191-206.

TROAKE, R.P. and WALLING, D.E. (1973). The natural history of Slapton Ley Nature Reserve VII: The hydrology of the Slapton Wood stream: a preliminary report. *Field Studies* 3, 719-740.

VAN VLYMEN, C.D. (1980). The natural history of the Slapton Ley Nature Reserve XIII: the water balance of Slapton Ley. *Field Studies* 5, 59-84.

Chapter 2
The geology and structure of the Slapton region, South Devon

Keith Chell
Field Studies Council
Slapton Ley Field Centre

The Devonian and Permian successions of the Slapton area reflect the principal events of the Variscan Orogeny. The fluvial shales and sandstones of the Lower Devonian, which are well exposed in coastal locations and form the predominant geological unit in the area, change to marine shales and limestones elsewhere in the county. Late Carboniferous tectonic activity, related to the main Variscan events, produced low grade regional metamorphism throughout the area, with the exception of the extreme south (Bolt Tail to Start Point). This zone reflects higher grade metamorphism, with extensive outcrops of mica schist, and hornblende chlorite schist. The Middle and Upper Devonian succession, and the Carboniferous succession are missing from this area, and reflects Variscan uplift and erosion. The final phase of the orogenic cycle is represented by 'red-bed' deposition of Permian age. These terrestrial red sandstones and breccias comprise small outliers in Slapton village (823448) and, 15Km to the west, at Thurlestone (678419). They rest uncomformably on older rocks and are undoubtedly the remnants of a far more extensive unit, which is more fully exposed in the Torbay area. A geological map of the Slapton region is shown in Figure 2.1.

Lower Devonian Slates

The Dartmouth Slate is the oldest formation recognised in the Slapton area. It comprises several hundred metres of red and green sandstones, siltstones and mudstones. Dineley (1966) has indicated the presence of fining-upward cycles within the formation, which compare closely with similar cycles in the Old Red Sandstone of the Welsh Borderland. Rare ostracoderm and placoderm fish fragments and plant remains suggest non-marine deposition. Pterapsid fish are poorly preserved in the finer units, and can be used to date the Dartmouth Slate to the Siegenian Stage of the Lower Devonian (Anderton *et al.*, 1979). The Dartmouth Slate is succeeded by the Meadfoot Slate. The Meadfoot slate indicates a stronger marine influence in the depositionary process. Mudstones, siltstones and sandstones with sporadic thin shell bands, linked with scour sole structures suggest deposition in a high-energy environment associated with a transgressive sea. Fossil remains date the Meadfoot Slate to Middle Siegenian/Emsian.

The overall palaeogeographic picture of the Lower Devonian in the area suggests fluvial deposition on the coastal plains of an Old Red Sandstone continental margin. A transgressive sea introduced the first Devonian Marine incursion - an onset of marine deposition which dominated until the end of the Devonian period.

2. The geology and structure of the Slapton region, South Devon

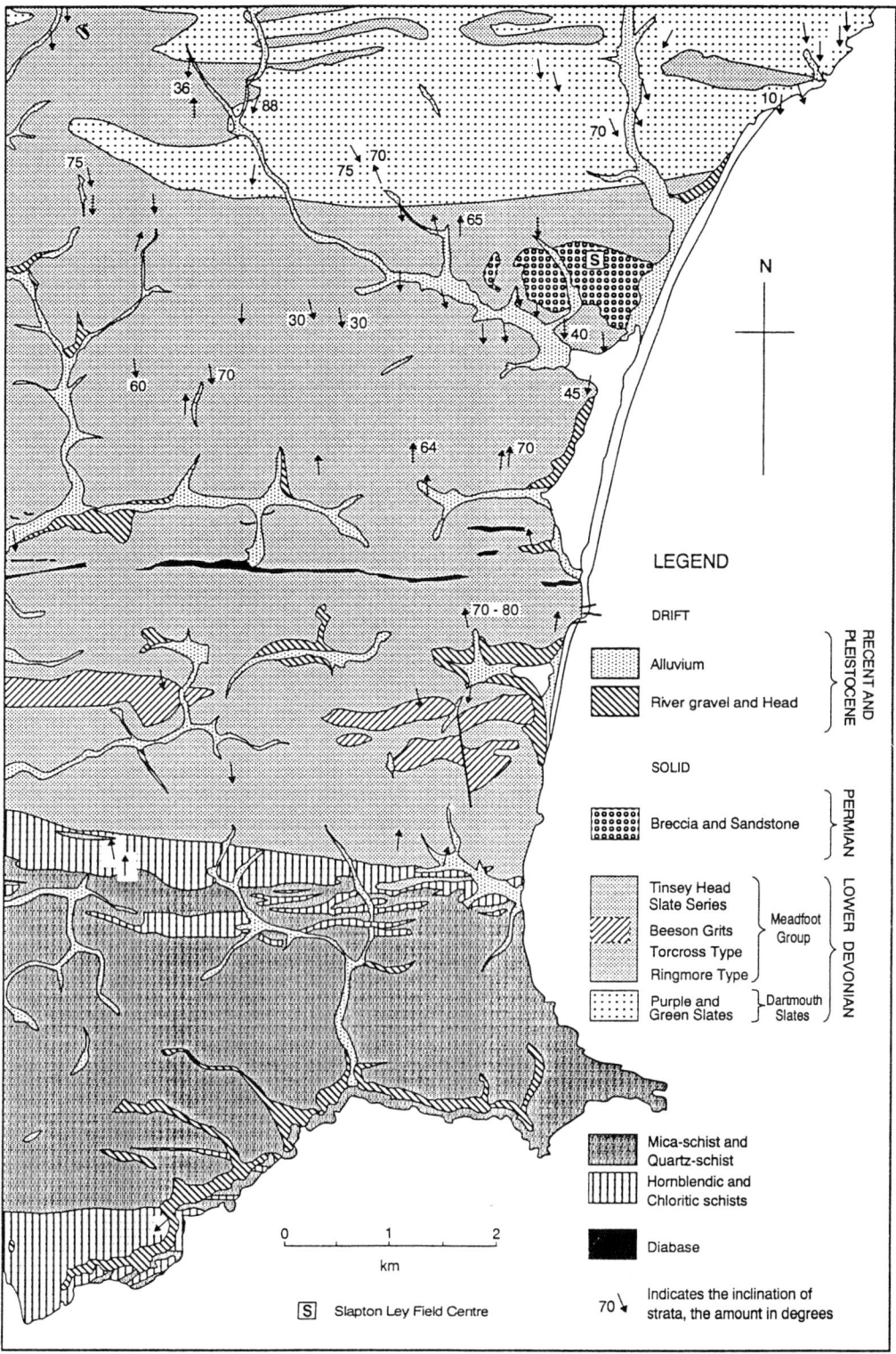

Figure 2.1 The geology of the Slapton region.

A field guide to the geomorphology of the Slapton region

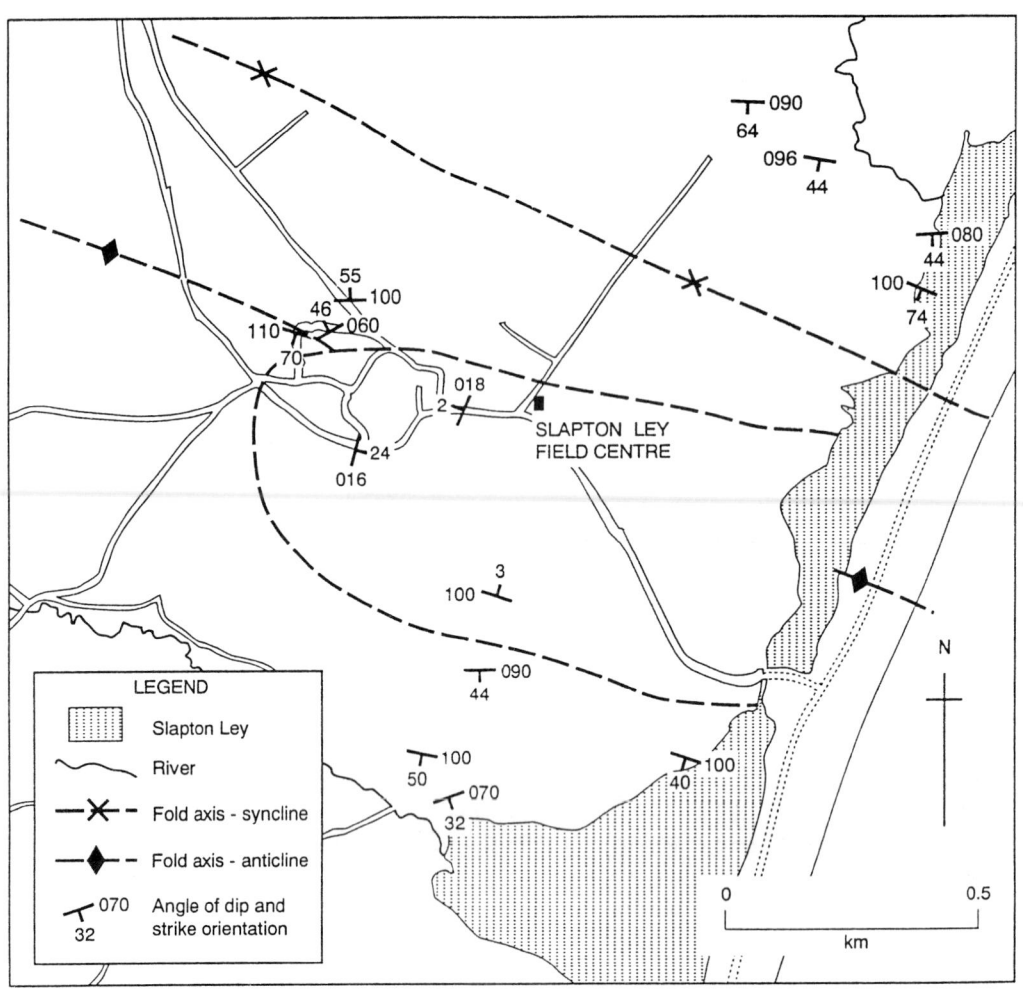

Figure 2.2 *Geological structures in the immediate vicinity of Slapton.*

Bolt Tail - Start Point Schist

The schist in the extreme south of the area is separated from the Meadfoot Slate by a steeply-angled east-west aligned boundary fault. Two main groups of schist are recognised - mica schist and hornblende-chlorite schist. This variation reflects the original pre-metamorphic material (possibly shale and tuff). The metamorphic event producing the schist has been radiometrically dated at 299Ma. This corresponds closely with similar work undertaken on the Meadfoot Slate (305Ma), suggesting that the schist is contemporaneous with the slate in terms of age of original, pre-metamorphic rock (Selwood *et al.*, 1982).

Permian Red-Beds

The red sandstone and breccia of the Slapton outlier rests uncomfortably on the Devonian Meadfoot Slate. The Sandstone comprises low-angled cross-bedded units, with depositionary structures in the breccia being more difficult to detect. The red-beds reflect the final stage of the Variscan Orogeny, with degradation of an upland area, and rapid infill of surrounding basins occurring under semi-arid conditions.

Geological Structure

The Variscan structures in the area are predominantly east-west trending (Figure 2.2). Polyphase deformation occurred, with evidence for F_4 folding present in the form of flat-lying kink bands of the Meadfoot Beds. The first phase of deformation produced the important folding, however, with F_1 folds lying broadly east-west. Axial planes of these F_1 folds are vertical in the Slapton area, but become less steeply inclined both north and south, giving a broadly anticlinorial structure. (Selwood *et al.*, 1982).

Minor Igneous intrusions

A series of small (<1m wide) dolerite dykes outcrop at Limpet Rocks, Torcross (823416) can be traced westwards across the area. The dykes follow an early cleavage pattern within the Meadfoot Slate, and are clearly post Lower Devonian.

REFERENCES

ANDERTON, R., et al., (1979) *A Dynamic Stratigraphy of the British Isles* Allen & Unwin

DINELEY, D.L., (1966) The Dartmouth Beds of Bigbury Bay, South Devon Quarterly Journal of the Geological Society 122 187-217

SELWOOD, E.B., et al., (1982) 'The Variscan Structures' in the Geology of Devon Ed DURRANCE and LAMING, University of Exeter.

Chapter 3
Weathering and soils

Steve Trudgill
Department of Geography
University of Sheffield

Weathering and soil formation

The bedrock of the Slapton area is a lower Devonian slate, with local outcrops of Permian deposits (see Chapter 2, this volume). The slates are approximately 390 million years old and originally composed of muds of low nutrient content. The muds, having been compressed and metamorphosed into slates, are dominated by quartz and kaolinite minerals. These are chemically very stable and weathering releases only small amounts of silica and aluminium and very little in the way of important plant nutrients such as calcium and magnesium. Iron is, however, a significant constituent of the rock; in the unweathered rock is mostly present in the greyish-coloured reduced (Iron II) form. During weathering this oxidises to the reddish oxidised (Iron III) form which give the soils their characteristic colour: the red colour of the soils is also thought to have been considerably enhanced during strongly oxidising weathering conditions which occurred during the Triassic period.

During the Pleistocene period, south Devon was not covered by ice but was at the margins of the ice sheets - as deduced by evidence of glacial drift deposits in north Somerset and along the southern coast of the Bristol Channel. Weathered deposits were not, therefore, moved by glacial erosion but by periglacial processes (see also Chapter 8, this volume). This has resulted in a thin soils on the slope crests, with thicker soils developed on periglacial deposits at the slope foot. Here, the regolith can be several metres deep, giving a permeable subsoil of soliflucted rock fragments. On the plateaux, the soil remains largely intact with soils up to about 1 - 2m deep.

The slate weathering, apart from the chemical processes of the oxidation of iron and some breakdown of clay minerals, has thus been dominated by mechanical weathering processes, especially during periglacial conditions.

Soil characteristics

Comprised originally of silts, the slate rock breaks down to yield a soil with abundant silt-sized particles, about 0.02 - 0.002 mm in diameter. Clay is also present, derived from the parent rock or formed in the profile during weathering. Small and large stones of slate are also often abundant in the soil profile, with many slate fragments of sand size. Commonly, the soils have 30 - 40% silt and 30 - 40% clay.

The soils are dominantly acid, silty and of a reddish (oxidised) colour. In soil type, they range from an acid brown earth towards a more podzolic soil, especially on undisturbed, north facing woodland slopes - elsewhere cultivation has redistributed the surface organic matter characteristic of podzolic soils and liming acts to reduce the acidity. In the valley bottoms, alluvial deposits exist and these give rise to waterlogged, gleys soils of a grey colour where the soil is not oxidised and with some red-mottling where oxidation has occurred.

i. Brown earths and brown podzolic soils

The brown earth has a well-mixed 'A' horizon (mixed humus and mineral layer), mixed by cultivation on agricultural land and by biological activity under woodland. pH values can be as low as 4.5 but where liming has occurred, this may rise to pH 7. Below this is a reddish 'B' horizon passing to fragmented bedrock. A representative woodland profile is shown in Table 3.1.

Table 3.1 Brown earth under woodland.

Horizon	Depth (cm)	Description
L	5 - 3	Litter of *Castanea*; some *Pinus* needles. pH 5.5
F	3 - 2	
H	2 - 0	Dark humus, 5.7 YR 2.5/1, black. pH 4.5
Ah	0 - 20	5 YR 3/2 dark reddish brown. Weak, granular - subangular blocky, friable. Abundant small stones. pH 4.2
AB	20 - 25	5 YR 3/3 dark reddish brown. Weak granular-subangular blocky. Abundant small stones. pH 4.3.
B	26 - 65	5YR 5/7 yellowish. Weak granular-subangular blocky. Abundant small-medium stones. pH 4.4.
C	65+	7.5 YR 7/8 reddish yellow. Weathered angular slate with a soil matrix between. pH 4.2.

Soil properties can vary with land use, with woodland generally having higher organic matter (5 - 15%), lower bulk densities (around 1.0 gm cm^{-3}) and higher infiltration rates (200 - 1500 mm hr^{-1}) in the topsoil than agricultural land. The agricultural land varies considerably in its properties according to crop type and time of cultivation. Organic matter is generally higher in pasture than in arable soils, bulk density is normally higher in arable soil than in pasture although infiltration rates can be very high in arable land if recently cultivated (up to 1500 mm hr^{-1}, or more) if recently tilled, or very low (down to 10 -20 mm hr^{-1} and even close to zero) if the surface soil has been exposed to rain splash, with consequent crusting, and/or compaction.

A property of considerable interest has been soil nitrate values. These can vary considerably with season and with land use (Trudgill, *et al.*, 1991). Generally the nitrate values in woodland soils are sustained by the mineralisation of organic matter in a warm, moist climate and the levels here can be higher than on agricultural land if the latter has not received fertiliser within 2 - 6 months prior to sampling. Subsequent to fertiliser application, the order is commonly arable > pasture > wood, reflecting the relative amounts of fertiliser applied.

Within Slapton Wood, the north-facing slopes tend to exhibit an accumulation of darker coloured organic matter accumulation than on the south-facing slopes. On these north-facing slopes, there are signs of incipient podzolisation in the soil, with brighter redder colours in the 'B' horizons indicating greater iron oxide content. Thus the more characteristic woodland brown earth has a browner humus and 'B' horizon colours of dominantly 10 YR or 7.5 YR and occasionally 5 YR while the woodland soils have a blacker humus and a 'B' horizon colour dominantly of 5 YR and occasionally 2.5 YR. Their disposition is shown in Trudgill (1983).

Agricultural soils tend to exhibit more intimate mixing of organic with mineral matter and are often consequently less red, commonly 7.5 YR with some 10 YR colours. However, redder, sandier soils occur on Permian breccia around Slapton village.

ii. Gley Soils

These are found in the valley bottoms on alluvial deposits and can have a silty clay texture and are thus finer textured than the silt loam brown earths. A representative profile is described in Table 3.2.

REFERENCES

TRUDGILL, S.T., 1983. The soils of Slapton Wood. *Field Studies* 5 (5), 833 - 840.

TRUDGILL, S.T., BURT, T.P., HEATHWAITE, A.L, and ARKELL, B.P., 1991. Soil nitrate sources and nitrate leaching losses, Slapton, South Devon. *Soil Use and Management* 7 (4), 200 - 206.

Table 3.2 Gley soil.

Horizon	Depth (cm)	Description
L, F, H absent due to trampling and effects of winter flooding.		
A(g)	0 - 15	5 YR 4/4 reddish brown. Gritty silt loam. Coarse subangular blocky-fine moderately developed structures. Few fine fissures cracking to surface. * 2% mottles 5 YR 4/6 yellowish red disposed along fissures. Firm consistence. pH 5.6
B(g)	15 - 45	7.5 YR 5/4 brown. Silt loam. Coarse subangular blocky, moderately developed structure. Few fine pores, 0.1%. Few medium/fine woody roots. Moderately firm consistence. 5% mottles 5 YR 4/6 yellowish red. pH 6.1.
C1(g)	45 - 55	10 YR 6/2 light brownish grey. Silty clay/silt loam. Weak consistence. Fine pores, iron stained, 10%. 15 - 20% mottled, 5 YR yellowish red. weak medium subangular blocky structure. pH 5.6.
C2(g)	55 - 60	5 Y 5/1 grey unmottled blue-grey silty clay. pH 5.2.
C3	60+	Gravel, small platy particles. Standing water.

* Profile surveyed in July, cracking to surface is absent or less marked in winter.

Chapter 4
Hillslope hydrology and geomorphology

Tim Burt

School of Geography
Oxford University

As noted in Chapter 1, the hydrology of catchments in the Slapton region is dominated by subsurface runoff. Only 1% of the annual runoff in the Slapton Wood stream is quickflow (Troake and Walling, 1973). Not all of this is surface runoff: some is rapid subsurface flow from saturated zones close to the stream, or via macropores. The remaining 99% is entirely generated as subsurface flow from within the soil and the fractured, upper part of the bedrock.

The infiltration capacity of soils in the Slapton area is naturally high; only when agricultural mismanagement has happened does the infiltration capacity fall to a level where overland flow can occur in heavy rain. The silty loam soils, normally freely draining, are easily compacted. Even light use of soils reduces the infiltration capacity below that of woodland or freshly ploughed soil (Table 4.1). Some agricultural practices, such as rolling to produce a fine tilth for optimal operation of herbicides, may have a particularly deleterious effect, reducing infiltration capacity to the point where even light rainfall can produce infiltration-excess overland flow. It is thought, however, that surface runoff has been rare in the Slapton region until the last decade (Heathwaite and Burt, 1992) and that surface erosion has not been instrumental in the formation of hillslopes in the area.

The main bulk of the shale bedrock is totally impermeable, but the soil and regolith are very permeable, allowing large volumes of subsurface stormflow to be produced quickly after rainfall. Hydraulic conductivities of up to 1 m/hr have been measured at the base of the soil profile (Burt and Butcher, 1985). Although close to the surface the hydraulic conductivity may be reduced because of compaction, it is likely to exceed 50 mm/hr throughout most of the profile. Much of the throughflow occurs in winter in the form of delayed hydrographs; discharge peaks several days after the rainfall, with high flows lasting for as long as two weeks. The delayed hydrographs occur only in the winter months when the soil moisture deficit has been recharged so that percolation into the regolith below can recommence. Burt and Butcher (1985) examined 44 such hydrographs for the

Table 4.1 *The effect of land use on surface runoff from hillslope plots.*
Results obtained using a rainfall simulator except where indicated.

Land use	Rainfall intensity (mm/hr)	Infiltration capacity (mm/hr)	Bulk density (g/cm^3)
Temporary grass	12.50	12.33	0.96
Barley	12.50	11.04	1.08
Rolled, bare ground	12.50	4.00	0.93
Lightly grazed, permanent pasture	12.50	5.85	1.12
Heavily grazed, permanent pasture	12.50	0.10	1.18
Permanent pasture (*)	-	3 - 36	-
Freshly ploughed soil (*)	-	50	-
Woodland soil (*)	-	180	-

(*) Determined using a ring infiltrometer

A field guide to the geomorphology of the Slapton region

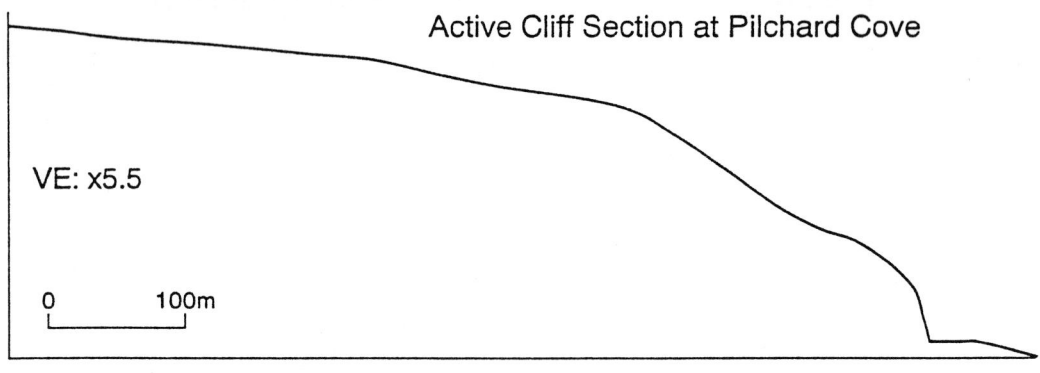

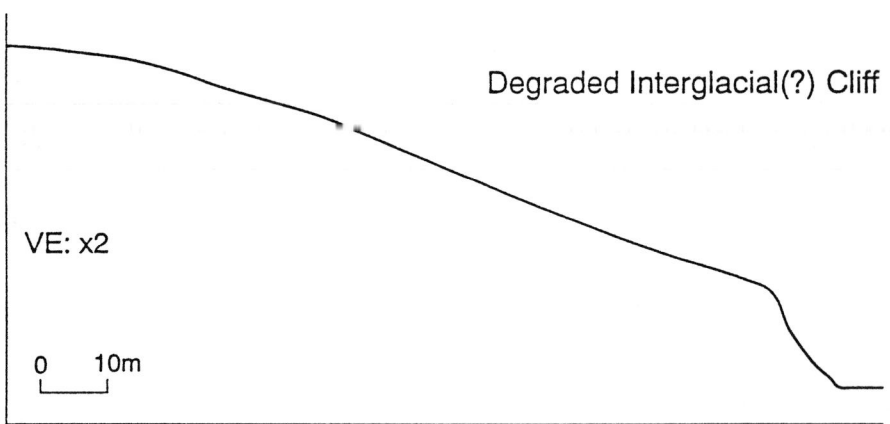

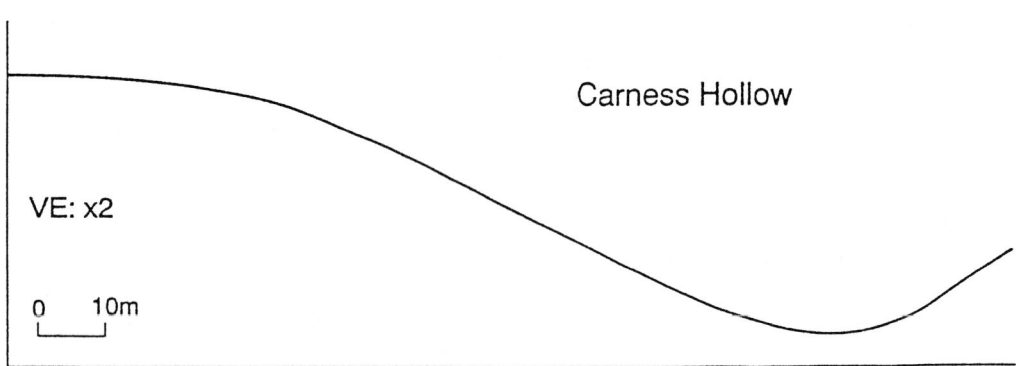

Figure 4.1 *Typical slope profiles from the Slapton region.*

period 1969-82. They found that during the winter half of the year, these hydrographs accounted for 46% of the runoff and occupied over half the period; by contrast quickflow occupied only 2% of the time and produced less than 4% of the runoff. The generation of the delayed hydrograph is associated with the formation of a large "wedge" of saturated soil which builds up within the soil and regolith above the impermeable bedrock. The saturated wedge extends upslope from the stream and, during periods of high subsurface discharge, it includes large parts of the plateau area above the valley-side slopes (Burt and Butcher, 1985). Drainage from the plateau is thought to be particularly important in generating the delayed hydrograph; much of this drainage takes place through large hillslope hollows. The delayed hydrograph is an especially important time for nitrate leaching since both flow and concentration are high at that time; it is also an important time for solute leaching in general (Burt et al., 1983; see also Chapter 5).

The shale bedrock is relatively solute-poor (Trudgill, this volume) so that chemical denudation is not thought to have been a dominant process in the shaping of hillslope forms in the Slapton area. It is possible that leaching rates may decline downslope (cf Crabtree and Burt, 1983), leading to slope decline over time. The tendency for more acid soils to form on the interfluves lends some support to this idea. It is clear that chemical weathering has helped to produce the deep permeable soils which are so important to the subsurface hydrology. It is possible that such weathering has been more important on the schists of the Start Point area; this might help to account for the widespread solifluction features found on that part of the coast compared to the slate and shale zones further north (See Chapter 8).

Slopes in the Slapton region are predominantly convex in form. This suggests that creep and solifluction have been the main formative processes. Below the convex segment, a straight section is often present, especially in the deeper valleys. Basal concavities, if present, tend to form a very small proportion of the total slope length. Maximum slope angles on the Devonian shales reach no more than 25 degrees. In the absence of any shear strength measurements, one cannot be sure about the stability conditions prevailing on the straight slopes. However, following Carson and Petley (1970), it seems likely that the soil is stable with respect to saturated conditions. As the rivers downcut their valleys, saturated soil conditions would have occurred often enough to prevent steeper slopes from remaining stable, since such slopes can exist only if the soil remains unsaturated. In support of this argument, it is worth noting that the "straight" slope sections found within the Slapton Wood valley appear rather uneven (compared, say, to slopes found on the Chalk escarpment) which seems to indicate signs of mass movement.

A series of profiles in the Slapton area (Figure 4.1) may be used to demonstrate the evolution of slope form. Savigear (1952) reported a series of slope profiles along the coast of South Wales between Laugharne and Pendine; this is a frequently quoted example of successful space-time substitution. The site was chosen by Savigear because the growth of a spit along the coast from west to east meant that slopes at the eastern end have only recently been protected from undercutting, whereas those at the western end have been sheltered for a much longer period of time. Kirkby (1984) has used a computer simulation model to recreate the Savigear sequence. A similar series of slopes forms may be detected in Start Bay. Modern cliffs are cut into the base of much older convex slopes. Kirkby's simulations show that such large convexities could not have formed during the Holocene; at least 100,000 years of mainly periglacial climate would be needed for their development. Active cliffs cut into Devonian slates can be found at the north end of Start Bay, in Pilchard Cove. As one moves south towards the Ley, the cliffs appear more degraded; growth of the shingle ridge at the north end of the bay has protected the cliffs from marine erosion (Chapter 9). Mercer (1966) has identified a degraded cliff from the last interglacial on the western shore of the Lower Ley. This 35 degree slope is mantled by a stony soil (possibly akin to the taluvium of Carson and Petley, 1970) and appears to represent a slope form transitional between an active cliff and an inland valley slope. As noted above, valley-side slopes are mainly convex; their straight segments reach angles of 20 - 25 degrees. It is not clear whether these straight sections represent a stability threshold or, as Kirkby argues for the Savigear series, are simply the result of a shared history of slope development. Kirkby's model shows that, once basal incision ceases, the maximum slope angle declines very slowly over time such that slopes of 250,000 years and 1 millions years are very similar. If a threshold does exist, its influence will be to make this angle even more widespread.

Given the time needed to form the summit convexities, Kirkby concluded that the general relief between Pendine and Laugharne predates the last two glacial periods. A similar case can be made for the slope forms of the Slapton region; inland valleys may well be at least 500,000 years old. Today, fluvial and coastal processes act to modify what seems to be, in general terms at least, a Tertiary and Quaternary landscape. It is clear that climatic conditions have been very different at some stages in the past. However, it is not certain for just how long the cold periods lasted, or what conditions were like during the interglacials. It is possible that the fluvial landscape evolved largely under climatic conditions which were not so different from those prevailing today. If this were so, then the hillslope forms observed in the Slapton region might not necessarily be relict features inherited from the cold periods. However, if creep were responsible for the summit convexities (rather than more active solifluction) then the valleys might well have needed several million years in order to reach their current stage of development.

REFERENCES

BURT, T.P. & BUTCHER, D.P. (1985). Topographic controls of soil moisture distributions. *Journal of Soil Science* 36, 469-486.

BURT, T.P., BUTCHER, D.P., COLES, N. & THOMAS, A.D. (1983). The natural history of the Slapton Ley Nature Reserve XV. Hydrological processes in the Slapton Wood catchment. *Field Studies* 5, 731-752.

CARSON, M.A. & PETLEY, D.J. (1970). The existence of threshold slopes in the denudation of the landscape. *Transactions of the Institute of British Geographers* 49, 71-95.

HEATHWAITE, A.L. & BURT, T.P. (1992). *The evidence for past and present erosion in the Slapton catchment, southwest Devon*. In: M. BELL and J. BOARDMAN (eds), Past and Present Soil Erosion, Oxbow Monograph 22, 89-100.

KIRKBY, M.J. (1984). Modelling cliff development in South Wales: Savigear revisited. *Zeitschrift fur Geomorphologie* 28, 405-426.

MERCER, I.D. (1966). The natural history of the Slapton Ley Nature Reserve I: Introduction and morphological description. *Field Studies* 2, 385-405.

SAVIGEAR, R.A.G. (1952). Some observations on slope development in South Wales. *Transactions of the Institute of British Geographers* 18, 31-52.

Chapter 5
Sediment and solute dynamics

Tim Burt
School of geography
Oxford University

Louise Heathwaite
Department of Geography
Sheffield University

Sediment and solute yields measured at the basin outlet have often been used to assess losses from the catchment area. Such information has the clear advantage of providing estimates of average loss rates representative of sizeable areas and therefore the need for spatial sampling is avoided (Walling, 1990). However, much caution is needed when attempting to interpret yield data. Walling (1983, 1990) has emphasised the need to take into account the sediment delivery processes interposed between the point of erosion and the basin outlet. Perhaps only a small proportion of the soil eroded within a catchment will find its way to the basin outlet. Rather less attention has been paid to the idea of solute delivery. It might be thought that this would be an easier matter to understand than sediment delivery, given the relatively simple link between solute transport and water flow. However, the two topics may be more similar than we might first think: solutes, too, are subject to transformation and storage during their passage through the basin.

In the 1960s there was concern that Slapton Ley was becoming increasingly eutrophic. Accordingly, since 1969, a monitoring programme of weekly sampling has been maintained by Slapton Ley Field Centre to quantify runoff, sediment and solute outputs from the catchments into the lake. More recently, a number of experimental studies have sought to link runoff processes to the loss of sediment and nutrients from the catchment hillslopes. Though data exist for many solutes, most attention has been paid to nitrate and, to a lesser degree, phosphate. Soil erosion and sediment delivery have also been the subject of considerable attention (see also Chapter 6). In reviewing these studies, we emphasise the need for a scale-dependent approach which takes into account the sediment and solute delivery processes. However, given the preliminary stage of investigations (at Slapton and elsewhere), a large number of linkages in the delivery process remain poorly understood. The catchment areas studied are shown in Figure 5.1.

To quantify the generation of overland flow, suspended sediment and nutrients from different land uses, a series of experiments were conducted using a rainfall simulator (Heathwaite *et al.*, 1990 a, b). The results (Table 5.1) show that surface runoff from heavily grazed permanent pasture was twice that from lightly grazed areas and over ten times that from ungrazed temporary grass. Large amounts of runoff were also produced from soil which had been compacted by rolling. The largest losses of sediment and nutrients were from the heavily grazed pasture; nutrient losses from lightly grazed grassland were much less. In both cases losses of phosphorus (P) were mainly organic while nitrogen (N) was lost mainly as ammonium. There was much less runoff from the ungrazed temporary grass and from the cereal field; in both cases a much larger fraction of P was lost in inorganic form. There was a large loss of sediment from the bare ground but nutrient losses were quite low; N was lost in roughly equal amounts as ammonium and nitrate. Thus, heavily grazed grassland may be the source of high nitrogen, phosphorus and suspended sediment inputs to the stream system through surface runoff. Much depends on the infiltration rate associated with the particular land use at any one time and on the position of the field within the

A field guide to the geomorphology of the Slapton region

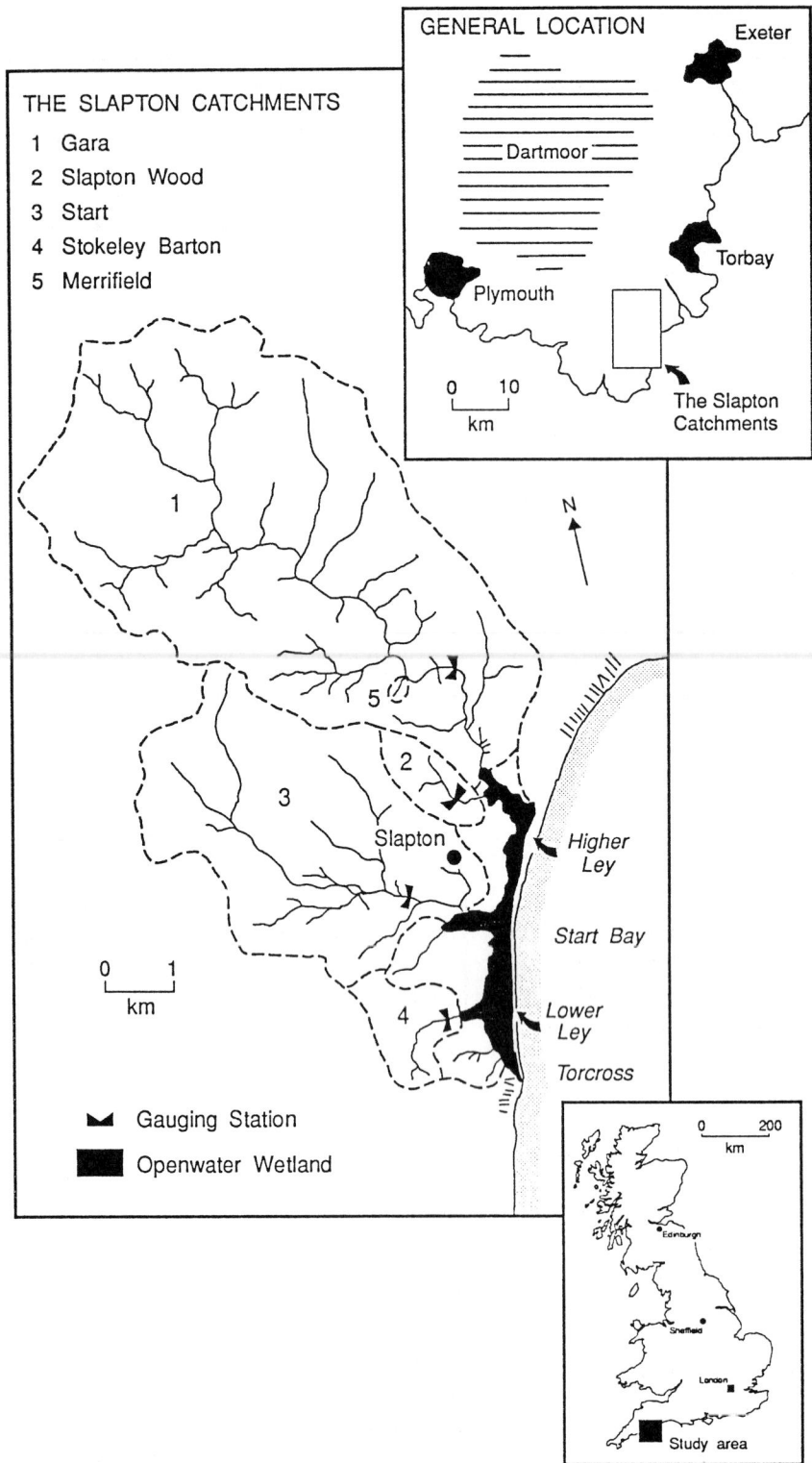

Figure 5.1 *The catchments draining into Slapton Ley.*

Table 5.1 *Runoff, sediment, nitrogen and phosphorus production from hillside plots.*

All rainfall simulation experiments lasted 4 hours at an intensity of 12.5 mm/hr. Infiltration data relating to these experiments are given in Table 4.1. nd : no data

	Heavily grazed permanent grass	Lightly grazed permanent grass	Temporary grass	Cereal (after harvest)	Bare ground (after rolling)
Total runoff (mm)	26.5	11.6	2.3	3.7	10.6
Sediment	22.28	0.37	0.15	0.31	5.10
Sediment per unit runoff (mg/mm)	840	31	65	84	481
Total P (mg)	124.63	3.33	0.73	0.76	1.87
Total P per unit runoff (mg/mm)	4.70	0.29	0.32	0.21	0.18
Inorganic P (mg)	21.45	2.00	0.33	0.35	0.49
Inorganic P per unit runoff (mg/mm)	0.81	0.17	0.14	0.09	0.05
Total N (mg)	69.83	2.96	nd	nd	3.40
Total N per unit runoff (mg/mm)	2.64	0.26	nd	nd	0.32
Inorganic N (mg)	64.15	0.81	nd	nd	0.82
Inorganic N per unit runoff (mg/mm)	2.42	0.07	nd	nd	0.08

catchment. There may be large losses from grazed land adjacent to a water course. Losses of P from bare ground are small, despite the large loss of sediment; even so, rates of soil erosion in the Slapton catchments are sufficiently high that soil-bound P may still be an important fraction of the total P loss.

There have been no lysimeter studies of nitrate leaching within the Slapton catchments. Soil profile studies of soil moisture and soil nitrate have been conducted using tensiometers, suction cup lysimeters and soil sampling (Trudgill *et al.*, 1991a). Coles and Trudgill (1985) used unbounded plots and ^{15}N to show that preferential flow of soil water down structural pathways can be responsible for the rapid movement of a proportion of surface applied nitrate fertiliser to soil drainage waters.

Burt and Arkell (1987) showed that all parts of the Slapton Wood catchment are significant non-point sources of discharge and nitrate, but hillslope hollows are particularly important and function effectively as point sources. Nitrate losses from different parts of the catchment were closely related to land use:

Headwaters
(mainly arable): 48.41 kg/ha/year

Valley-side slopes
(arable and grass): 44.56 kg/ha/year

Carness hollow
(arable and grass): 33.17 kg/ha/year

Eastergrounds hollow
(grass): 31.63 kg/ha/year

Slapton Wood
(woodland): 23.91 kg/ha/year

However, Trudgill et al., (1991a) could not establish clear relationships between soil nitrate content and land use or between soil nitrate and leaching loss. Leaching losses appear to relate more to the generation of subsurface flow than to land use alone. These results suggest that a scale-dependent approach is required in future studies which combines soil profile, plot and hillslope scale observations. It is difficult to relate soil conditions to patterns of leaching from a hillslope and to infer leaching mechanisms from observations made at the foot of a slope.

Table 5.2 Slapton catchments: land use and nutrient loads for the 1988 water year.

	Gara	Slapton Wood	Start	Stokeley Barton
Area (ha)	2362	93	1097	153
Runoff (mm)	920	581	950	148
Land use (%):				
Grass	81.2	32.1	52.6	28.8
Arable	11.9	36.1	34.2	66.0
Stream loads (kg/ha) - total load (t) is given below in brackets:				
Ammonium-N	1.79 (4.2)	0.16 (0.02)	1.39 (1.5)	0.23 (0.03)
Nitrate-N	68.03 (160)	63.60 (5.9)	103.93 (112)	19.51 (3.0)
Phosphate-P	0.38 (0.9)	0.21 (0.02)	0.58 (0.6)	0.21 (0.03)
Suspended sediment	503.34 (1190)	66.15 (6.2)	224.89 (242)	23.37 (3.6)

Table 5.2 shows the inorganic stream load for the 4 major catchments draining into Slapton Ley for the 1988 water year. These data were computed from continuous discharge observations and from water samples taken at intervals ranging from 15 minutes during storm events to a maximum of 24 hours. For the two intensively farmed catchments, Gara and Start, high loads are shown for all determinands. The Gara has a particularly high suspended sediment load which may be related to its steep slopes; high losses of P and ammonium suggest that grazing may be an important factor there. The Start has a high nitrate load which may relate to the greater amount of arable land in that catchment; ammonium and P loads are again high. Unfortunately, because of experimental difficulties, measurements of total N and P losses were not available.

There has been continued interest in temporal variations in stream load since Troake and Walling published their preliminary findings in 1973. Hydrographs and a nitrate chemograph for the Slapton Wood catchment are shown in Figure 5.2 for a period of high subsurface runoff. The main basin hydrograph shows clear discharge peaks associated with the rainfall; this indicates the generation of quickflow. Much of this quickflow will be overland flow from impermeable surfaces (tracks, roads, fields of low infiltration capacity) and from saturated areas, but some of the quickflow is rapid subsurface flow. This can be inferred from the behaviour of nitrate during the quickflow period; although there is some dilution (because surface runoff is low in nitrate), this effect is small. Given the large increase in discharge, this suggests that there must be a significant contribution to the quickflow hydrograph from 'old' subsurface water with a high nitrate concentration. The occurrence of delayed discharge responses

5. Sediment and solute dynamics

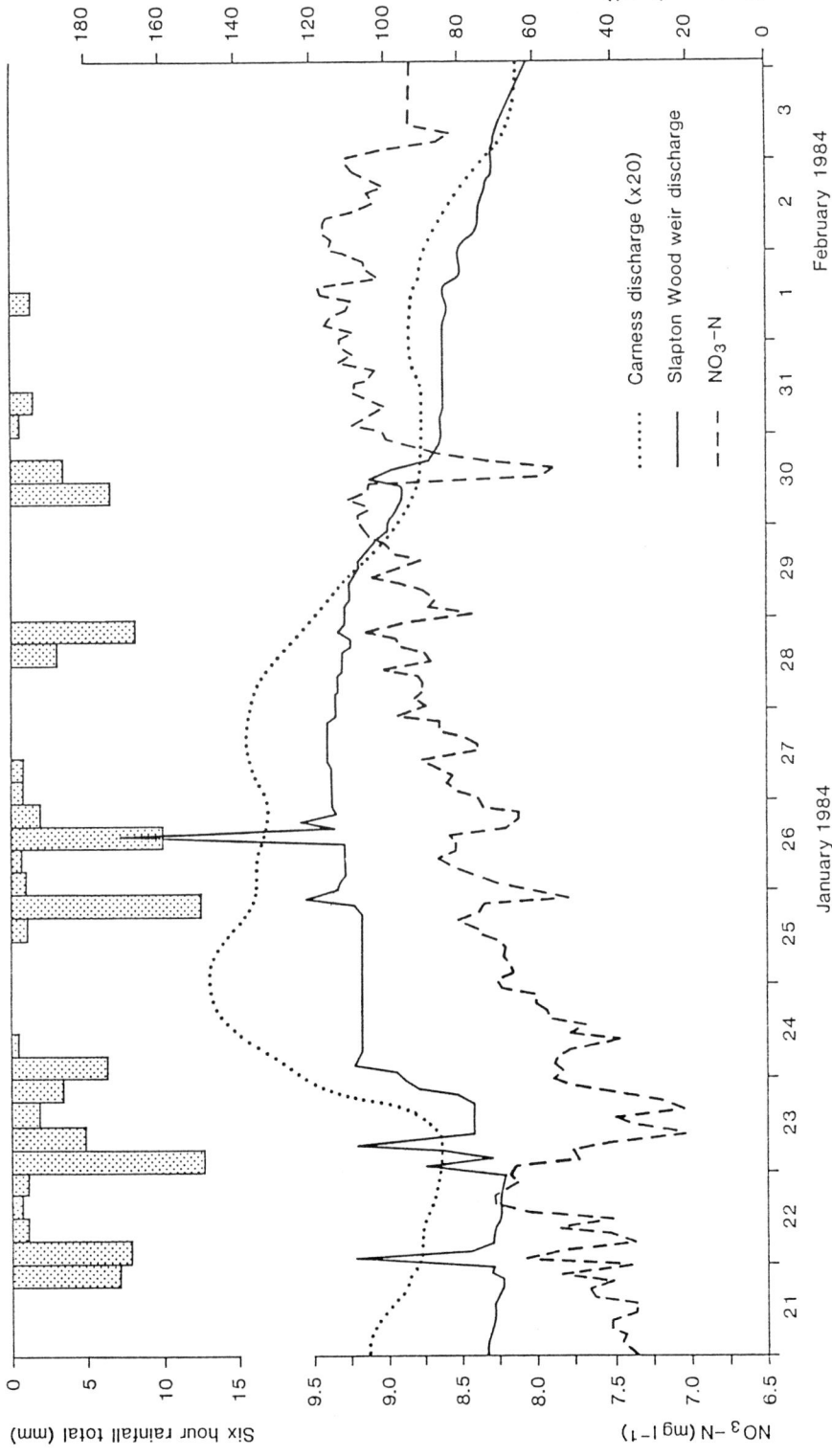

Figure 5.2 *Hydrographs and a nitrate chemograph for the Slapton Wood catchment during the 1984 winter.*

(see Chapter 4) is shown well by the hydrograph from the Carness subcatchment; this area lacks sources of overland flow and so the hydrograph is much more smooth. Burt et al., (1983) found that the delayed discharge peaks are strongly associated with nitrate leaching since both flow and concentration are high at such times (See also Burt and Arkell, 1987). In this example the peak nitrate levels occur during the flow recession, but on other occasions the delayed discharge and nitrate concentrations peak simultaneously. The nitrate load is highly dependent on discharge, particularly because the variation in nitrate concentration is relatively small.

Figure 5.3 shows the discharge response for the river Gara for a storm of 34.3 mm in March 1988. Here the influence of surface runoff is dominant since 13% of the rainfall is returned as quickflow. Suspended sediment concentration shows a typical response, increasing by several orders of magnitude from very low levels during baseflow (c 10 mg/l) to a peak of over 700 mg/l on the rising limb of the flood hydrograph. Phosphate concentrations follow a similar pattern, probably because the it has originated from the sediment. Ammonium concentrations also peak during the quickflow, though less dramatically. Like phosphate, the ammonium load is strongly dependent upon the quickflow discharge. Nitrate levels are quite strongly diluted during the quickflow; even so, the nitrate load clearly increases. A delayed discharge peak occurs about three days after the rainfall and there is a significant and protracted subsurface discharge response well above pre-strom levels. As would be expected, both nitrate concentration and load are high at this time.

The majority of the annual stream load is delivered in winter when discharge is high; usually loads peak in January and February. In the 1988 water year, 64% of the nitrate load in the Gara and 54% of the nitrate load in the Start were carried in these two months. Nitrate concentrations on the Start exceeded the EC limit (11.3

Table 5.3 Monthly totals of discharge, nitrate concentrations and nitrate load for the Slapton Wood catchment for the 1984 water year.

Month	Rainfall (mm)	Discharge (x1000 m^3)	Mean NO$_3$-N concentration (mg/l)	Nitrate load (Kg N)
October (1983)	75	11.28	6.37	71.86
November	69	11.83	7.00	82.25
December	117	77.99	7.75	604.72
January (1984)	243	172.93	7.79	1346.74
February	70	97.79	9.15	894.58
March	65	27.24	8.29	225.71
April	8	23.18	7.05	163.44
May	62	15.19	6.62	100.63
June	9	7.90	6.19	48.88
July	34	5.17	5.86	30.28
August	54	4.96	6.01	29.83
September	89	3.90	7.17	27.95
Totals	895	459.36	- **flow weighted mean =7.90**	3626.90

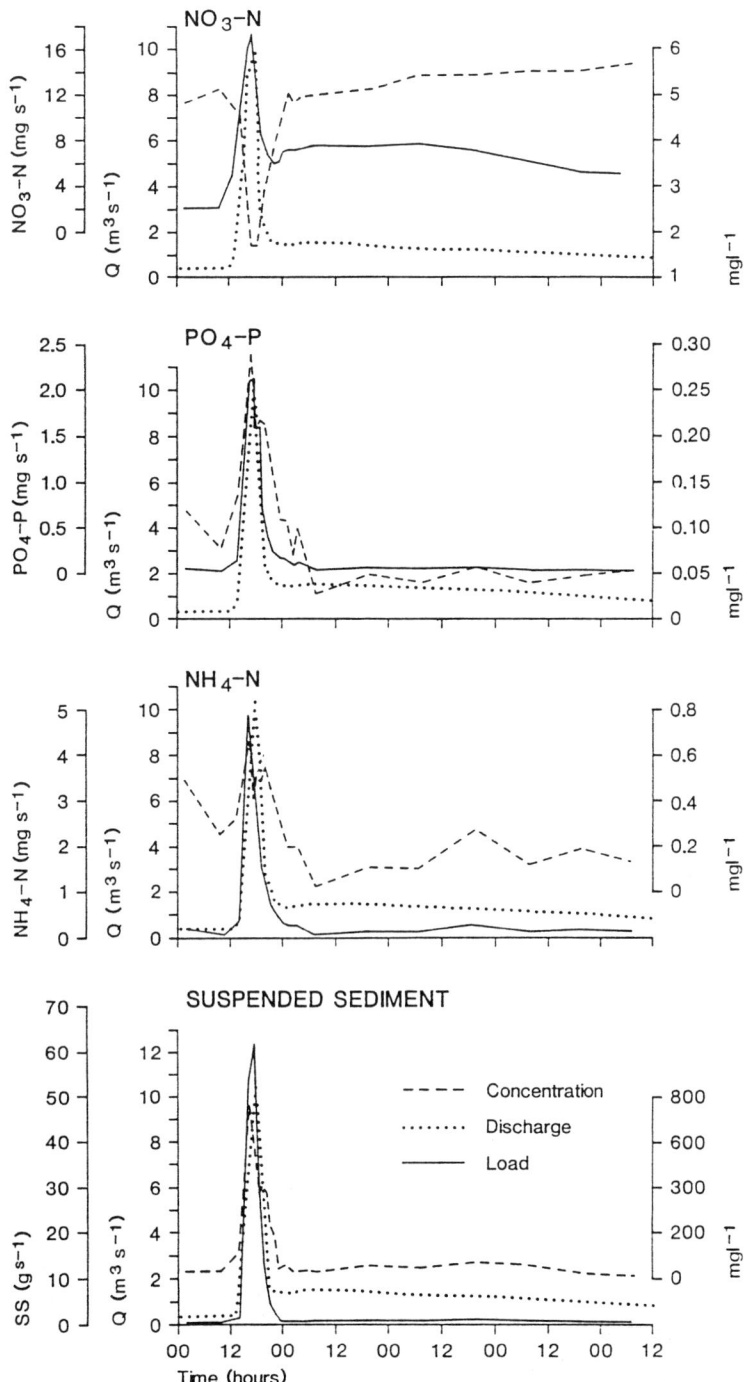

Figure 5.3 Hydrographs and chemographs for the Gara catchment for the storm of 23 March 1988.

mg/l NO_3-N) for four months. For suspended sediment, 93% and 70% of the Gara and Start load were delivered in January and February, a period when surface runoff production was at its maximum. The winter load of ammonium and P, being strongly related to the sediment transport, also peaked in those two months, about 70% of the annual load in both cases. Table 5.3 shows discharge and solute data for the Slapton Wood catchment for the 1984 water year. The seasonal pattern of flow and concentration is clearly evident. 78% of the nitrate load was lost between December and February. This was a relatively dry year and although nitrate concentrations were above 10 mg/l for much of February, at no time did they exceed the EC limit. By contrast, in the next winter which followed a protracted drought, nitrate concentrations were higher; they exceeded the EC limit for 8 days at the catchment outlet and for 57 days at the point upstream where the stream leaves the farmland and flows into the wood.

Despite the existence of over 20 years of sediment and solute data for the Slapton catchments, only the nitrate record has been investigated in any detail (Figure 5.4). The first 15 years of the record were discussed by Burt et al., (1988); the trends they observed continue to be evident. Use of partial regression analysis has shown that the long-term increase in nitrate concentration is the dominant pattern in the annual series; the effect of climate, in that wetter years have higher concentrations, is less important. This increase has been related to changes in land use and farming practice (Heathwaite and Burt, 1991). An important lag effect has also been identified in that dry years tend to be followed by higher nitrate concentrations in the following year, and vice versa. High nitrate levels an autumn following a summer droughts are especially notable. It appears that much labile organic matter may build up in the soil during hot, dry periods; this matter is easily mineralised as soils rewet and the resulting nitrate is vulnerable to leaching if the winter is a wet one.

The Higher Ley is a sediment trap for material eroded within the Gara and Slapton Wood catchments. Only Harris (1988) has measured sedimentation rates in the Higher Ley; further work is in progress. Using Caesium dating, Harris showed that sedimentation near the mouth of the Slapton Wood valley had been 25 cm in 40 years. Down-core variations in the magnetic properties of the sediment suggest that, over the last decade, hillslope sources have replaced channel sources as the dominant sediment source; this would accord with our own observation (Heathwaite et al., 1990a) that soil erosion has become a significant factor in the Slapton region in the 1980s. Heathwaite and O'Sullivan (1991) discuss the history of sedimentation in the Lower Ley and Owens (1990; see also Chapter 6) has examined deposition in the lower Start valley.

Heathwaite and Burt (1991) presented a monthly sediment and solute balance for the Higher Ley for the 1988 water year. The Higher Ley acts as an important sink for sediment and nutrients in winter, but functions as a source during the summer when flow is low and when pH/eH conditions at the sediment-water interface may favour release of nutrients, especially P (Trudgill et al., 1991b). For the 1988 water year as a whole, the Higher Ley was an important sink for ammonium, nitrate and suspended sediment, but a source of inorganic P. It may well be that the transformation of insoluble P, originally bound to soil particles, into more available forms is an important factor in the

5. Sediment and solute dynamics

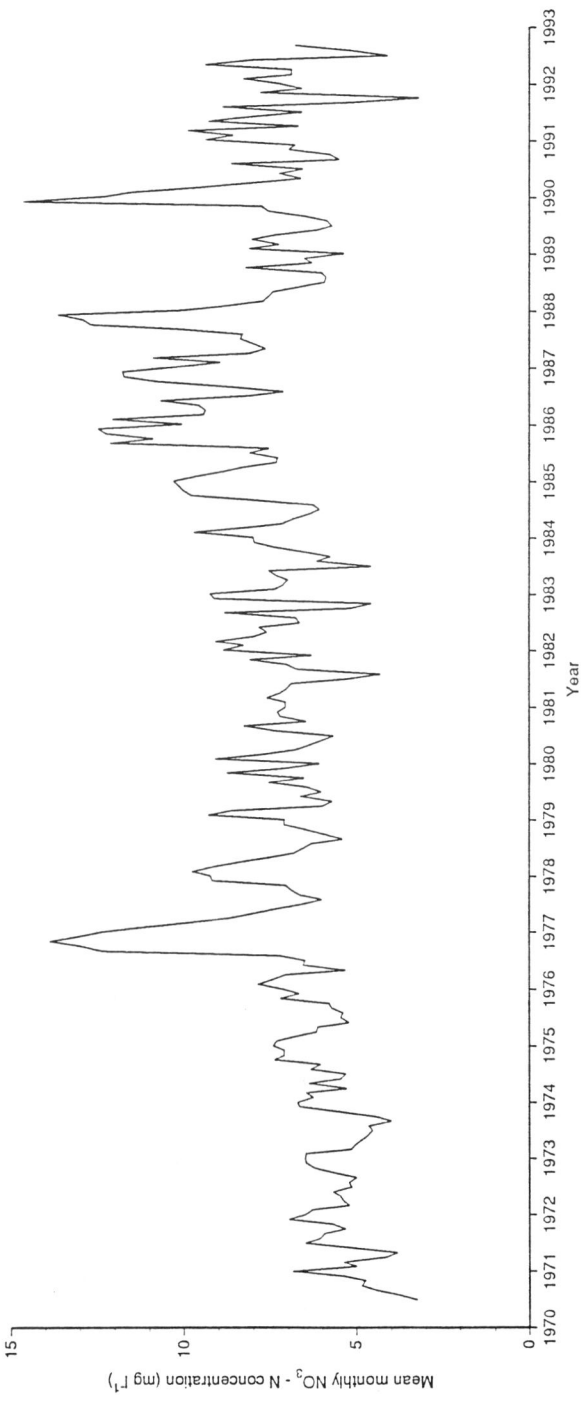

Figure 5.4 The long-term nitrate record for the Slapton Wood catchment.

Figure 5.4 The long-term nitrate record for the Slapton Wood catchment

eutrophication of the Ley. However, more work is needed to investigate the changing roles of N and P within the lake water through the summer season. On occasions, nitrate concentrations fall to very low levels; under such conditions, albeit temporary, N rather than P may be the limiting nutrient.

The Lower Ley has a flushing rate of 20 times per year (Van Vlymen, 1979), but in winter at peak flow, the lake volume is replaced every 3 days. This suggests that a large proportion of the stream load will be displaced from the lake in winter, thus having little impact on eutrophication. It is in summer, when the flushing rate is low, that stream inputs are most important; moreover, nutrient cycling within the lake water and the release of nutrients from lake sediments also becomes significant at this time. Work on N and P cycling within the lake system remains an important goal for future research.

References

BURT, T.P. & ARKELL, B.P. (1987). Temporal and spatial patterns of nitrate losses from an agricultural catchment. *Soil Use and Management* 3, 138-142.

BURT, T.P., BUTCHER, D.P., COLES, N. & THOMAS, A.D. (1983). The natural history of the Slapton Ley nature reserve. XV. Hydrological processes in the Slapton Wood catchment. *Field Studies* 5, 731-752.

BURT, T.P., ARKELL, B.P., TRUDGILL, S.T. & WALLING, D.E. (1988). Stream nitrate levels in a small catchment in south west England over a period of 15 years (1970-1985). *Hydrological Processes* 2, 267-284.

COLES, N. & TRUDGILL, S.T. (1985). The movement of nitrate fertiliser from the soil surface to drainage waters by preferential flow in weakly structured soils, Slapton, south Devon. Agriculture, *Ecosystems and Environment* 13, 241-259.

HARRIS, T.R.J. (1988). *Tracing sediment source linkages in a small catchment using mineral magnetism: the effects of land use change.* Unpublished B.A. Dissertation, School of Geography, Oxford University.

HEATHWAITE, A.L. & BURT, T.P. (1991). Predicting the effect of land use on stream water quality. *IAHS Publication* 203, 209-218.

HEATHWAITE, A.L. & O'SULLIVAN, P.E. (1991). Sequential inorganic chemical analysis of a core from Slapton Ley, Devon, UK. *Hydrobiologia* 214, 125-135.

HEATHWAITE, A.L., BURT, T.P. & TRUDGILL, S.T. (1990a). *Land use controls on sediment delivery in lowland agricultural catchments.* In: J. BOARDMAN, I.D.L. FOSTER & J.A. DEARING (eds), Soil Erosion on Agricultural land, Wiley, 69-97.

HEATHWAITE, A.L., BURT, T.P. & TRUDGILL, S.T. (1990b). *The effect of agricultural land use on nitrogen, phosphorus and suspended sediment delivery in a small catchment in south west England.* In: J.B. THORNES (ed), Vegetation and Erosion, Wiley, Chichester, 161-179.

OWENS, P.N. (1990). *Valley sedimentation at Slapton, south Devon, and its implications for the estimation of lake sediment-based erosion rates.* In: Soil Erosion on Agricultural Land, J. BOARDMAN, I.D.L FOSTER & J.A. DEARING (eds), Wiley, Chichester, 193-200.

TRUDGILL, S.T., HEATHWAITE, A.L. & BURT, T.P. (1991a). Soil nitrate sources and nitrate leaching losses, Slapton, South Devon. *Soil Use and Management* 7 (4), 200-206.

TRUDGILL, S.T., HEATHWAITE, A.L. & BURT, T.P. (1991b). The natural history of the Slapton Ley nature reserve. XIX. A preliminary study on the control of nitrate and phosphate pollution in wetlands. *Field Studies* 7, 731-742.

VANVLYMEN, C.D. (1979). The natural history of the Slapton Ley nature reserve. XIII: The water balance of Slapton Ley. *Field Studies* 5, 59-84.

WALLING, D.E. (1983). The sediment delivery problem. *Journal of Hydrology* 69, 209-237.

WALLING, D.E. (1990). *Linking the field to the river.* In: Soil Erosion on Agricultural Land, J. BOARDMAN, I.D.L. FOSTER & J.A. DEARING (eds), Wiley, Chichester, 129-152.

Chapter 6
Sediment yields and budgets in the Start valley

Ian Foster
Centre for Environmental Research and Consultancy
Coventry University

Des Walling
Phil Owens
Department of Geography
Exeter University

Introduction

Sediment yields measured from catchment monitoring programmes and/or reconstructed from lake sedimentation rates can be used to provide information on the impact of land use change on soil erosion. However, it is often difficult to demonstrate the exact link between land use and sediment yield since the yield of sediment from the drainage basin is a function, at least in part, of the sediment delivery ratio (Walling, 1988).

This brief review looks at various estimates of sediment yield in the Slapton Ley catchments and compares them with other estimates from the local region. With reference to radionuclide and mineral magnetic measurements on alluvial and colluvial deposits in the Start catchment, an attempt is made to identify sources and sinks of eroded sediment.

Sediment Yields

The most recent estimates of suspended sediment yield from the Gara, Slapton Wood, Start and Stokeley Barton streams were published by O'Sullivan *et al.*, (1989). Estimates range from under 10 to over 70 t km^{-2} yr^{-1} (Table 6.1). At Stokeley Barton, the yield is higher than either the Gara or Start streams, but the Slapton Wood yield is considerably higher than other calculated values (Table 6.1). The total sediment flux to the Lower Ley was calculated by O'Sullivan *et al.*, (1991) using ^{210}Pb and ^{137}Cs to derive a chronology and estimate rates of lake sedimentation. Around 8 t ha^{-1} of dry sediment are being deposited annually on the lake bed which is equivalent to a mass of 610 t or a sediment yield of 13.4 t km^{-2} yr^{-1}. These various studies suggest that a figure of ca. 10-20 t km^{-2} yr^{-1} is a reasonable estimate of sediment yield for the Slapton catchments. However, lake-sediment based reconstructions by Foster (1992) in the Old Mill catchment near Dartmouth has shown

Table 6.1 Estimates of Sediment Yield From the Slapton Catchments.

Catchment	Area[1] (km2)	Susp. Sed.[1] load (t)	Suspended Sediment load (t km^{-2} yr^{-1}) 1	2	3
Gara	23.62	218.48	9.25		
Slapton Wood	0.93	67.76	72.86	14.2[a]	8.4
Start	10.79	104.36	9.67		
Stokeley Barton	1.53	47.83	31.26	11.6[b]	
Total	36.87	642.3	17.42		

1. From O'Sullivan et al., 1989 (Data For 1987-88) 2. From Park, Pers Comm. a Based on four years of data 1978-1981 b Based on two years of data 1980-81 3. From Troake and Walling 1973

that sediment yields over the last 25 years, in an area of similar climate, lithology and catchment characteristics, lie between 80 and 100 t km^{-2} yr^{-1}; some 8 to ten times higher than the sediment flux to Slapton Ley (Figure 6.1). Three important questions therefore arise from the data presented above:

1. Do the differences between the Old Mill and Slapton sediment yields simply reflect differences in soil erosion rates between the two areas?

2. Are soil erosion rates similar but the differences controlled by a lower sediment delivery ratio in the Slapton catchments?

3. If the answer to question 2 is yes, what is the present location of stored sediment and where did it come from?

Sediment Concentrations in the Start stream

Several investigations have been undertaken in the lower Start valley (Figure 6.2) in order to observe and measure sediment transport processes and characterise sedimentary deposits. At high discharges in the lower 1.5 km of the Start valley downstream of Battleford, for example, sediment concentrations decrease by as much as 90% of their input level (Figure 6.3). Field observations by the authors suggest that these decreases are due to ponding upstream of Deer Bridge, causing sediment to settle out of suspension under reduced flow velocities. These results suggest that a considerable quantity of sediment is being lost from fluvial transport and diverted into floodplain storage.

The Start Valley Sediment Sinks

Surveys of the Start valley have identified that two areas form important sediment sinks; the valley bottom and upslope of field boundaries constructed approximately normal to the contour (transverse hedgerows). Surveys of the Start valley between 1986 and 1988 revealed extensive deposition of sediment between Deer Bridge and Battleford. Seven valley cross sections were surveyed by Owens (1990) who found uncompacted sediments to depths in excess of 1.5 m across the entire floodplain (Figure 6.4). The total volume of sediment stored in the valley was estimated to be 75,711 m^3 which is equivalent to 34,070 t of dry sediment. Soil depths on the south facing slopes of the lower Start Valley have been measured by augering along hillslope transects. Soil depths vary from as little as 9cm near the interfluves to as much as 94cm behind transverse hedgerows.

Sediment cores retrieved from the floodplain upstream of Deer Bridge and upslope from field boundaries have been analysed for the fallout radionuclides ^{137}Cs and ^{210}Pb in order to assemble information on sediment movement and storage within the Start catchment. Mineral magnetic properties have been measured in order to identify sediment source linkages.

Fallout Radionuclides in sediment budget investigations

Figure 6.5A illustrates a typical ^{137}Cs profile associated with a core collected from the floodplain of the Start Stream. Both the total ^{137}Cs inventory (790 mBq cm^{-2}) and the profile shape indicate that there has been substantial deposition of fine grained sediment since the early 1950s. The buried peak at 38cm below the surface indicates that around 38 cm of deposition has occurred since 1963. The ^{210}Pb profile which is also plotted for this core indicates that of the order of 80cm of deposition has occurred in the relatively recent past (ie ca. 100 years). More detailed interpretation of

Sediment yields and budgets in the Start valley

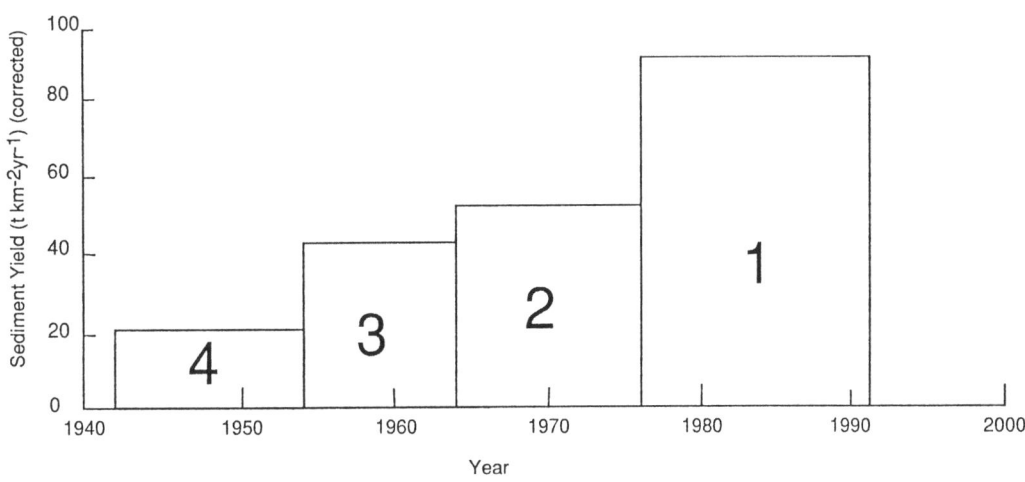

Figure 6.1 *Reservoir-sediment based estimates of suspended sediment yield in the Old Mill catchment, Dartmouth.*

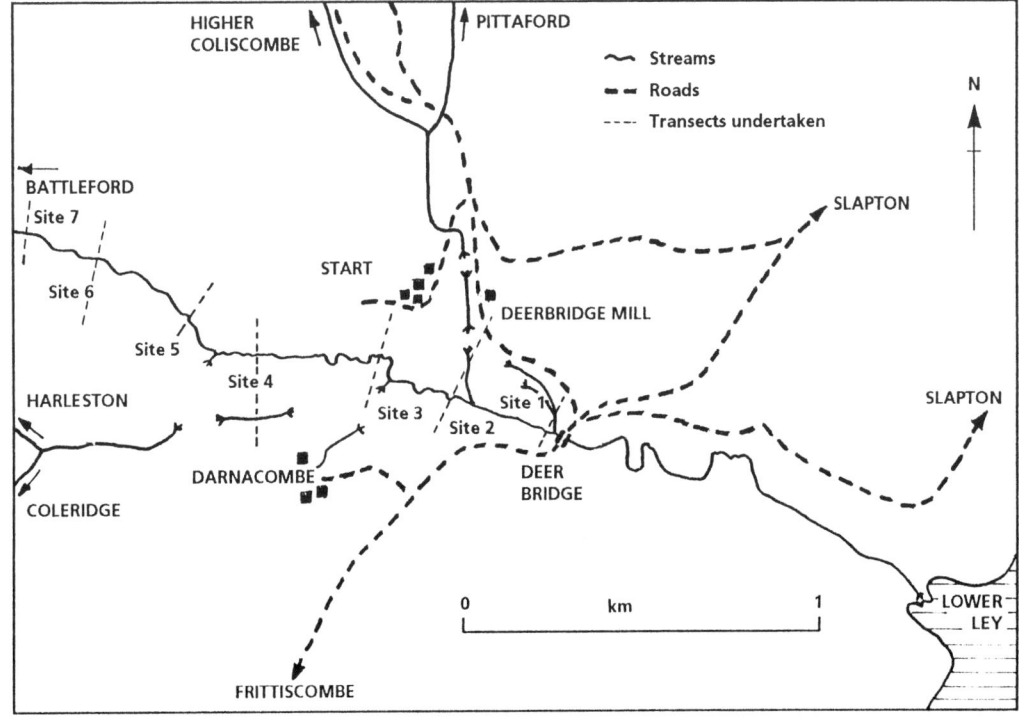

Figure 6.2 *The Lower Start Valley with transect locations. From Owens (1990).*

A field guide to the geomorphology of the Slapton region

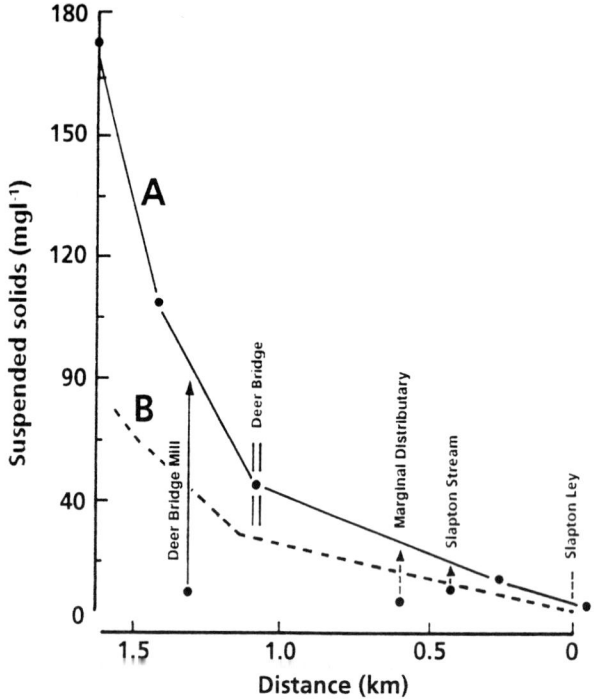

Figure 6.3 *Transmission losses of suspended solids through the Start Stream from Start to Slapton Ley A. From Foster (1986) Unpublished. B. From Owens (1990).*

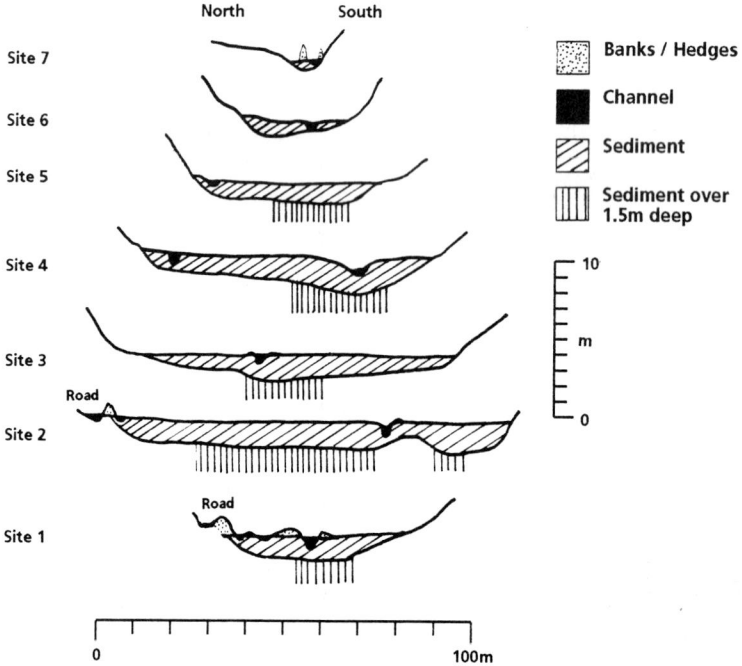

Figure 6.4 *Sediment accumulation in the Lower Start Valley. From Owens (1990).*

Sediment yields and budgets in the Start valley

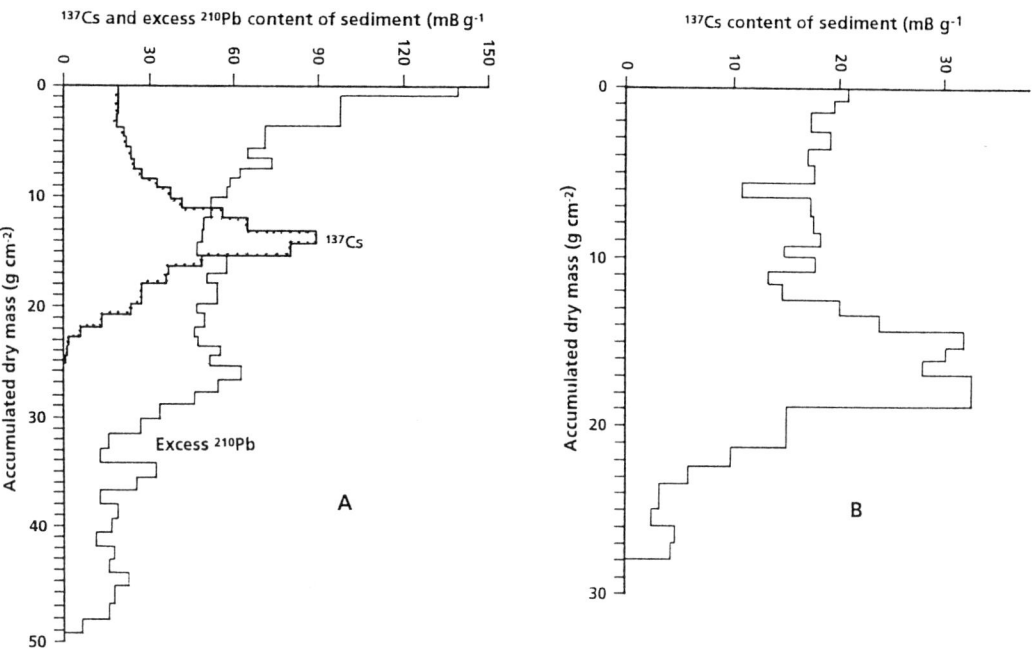

Figure 6.5 A *^{210}Pb and ^{137}Cs profiles from a valley bottom core (A) and a hedge boundry core (B).*

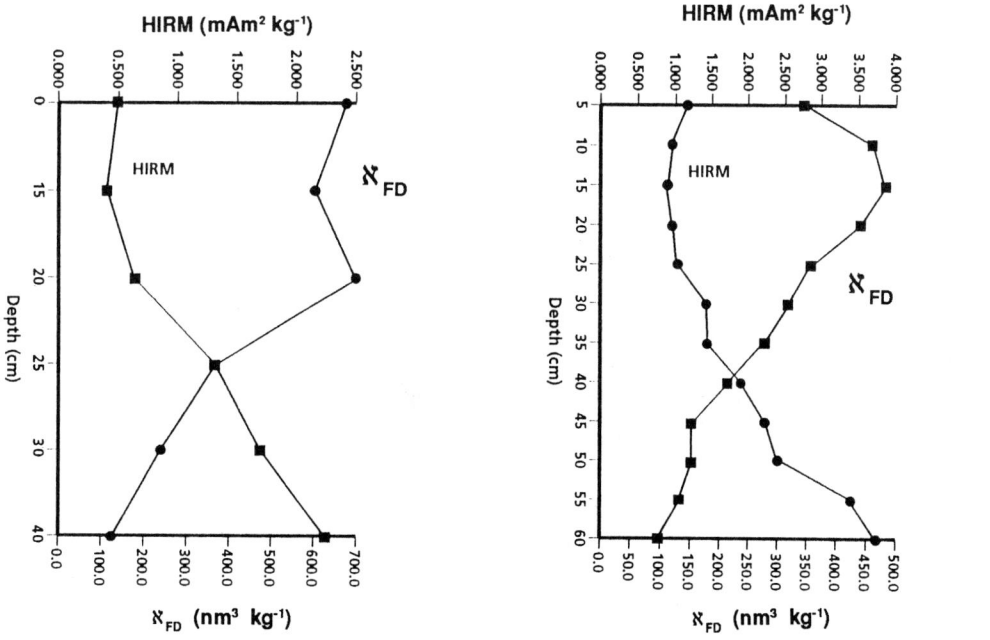

Figure 6.5 B *Frequency dependent susceptibility and HIRM traces in well drained arable (A) and Woodland (B) soils measured on sub 63 µm fraction of soil.*

the profile shape is currently being undertaken, but this suggests that rapid sedimentation occurred within the portion of the profile lying between 35 and 55 cm from the surface and this can be tentatively ascribed to wartime changes in land use. Estimates of sedimentation rates from dated cores by Owens (1990) indicate that between 341 and 1136 t of sediment were annually deposited in the flood plain in the recent historical past, increasing estimates of sediment yield to the lower Start valley to between 82 and 148 t km^{-2} yr^{-1}.

Cores collected from fields known to have been undisturbed for the past 50 years have provided a ^{137}Cs reference inventory value for the Start catchment of 261 mBq cm^{-2}. Figure 6.5 B illustrates a typical ^{137}Cs profile from a sediment sink located upslope of a hedge boundary at the base of a steep pasture field adjacent to a tributary of the Start stream. As with the floodplain core, both the total inventory and the shape of the profile indicate that a substantial amount of of deposition has occurred behind this boundary over the last 35 years. The buried ^{137}Cs peak, which is located some 15cm below the present surface, indicates that about 15cm of deposition has occurred since 1963.

Mineral Magnetism and Sediment Sources

Various attempts have been made to discriminate sediment sources from mineral magnetic studies of sources (topsoils, subsoils, channel banks, etc) and sediment sinks (colluvium, alluvium and lake sediments) (*e.g.* O'Sullivan *et al.*, 1991; Foster, 1992). These studies have shown that secondary magnetic minerals are often enhanced in topsoils, except in heavily gleyed soils, and that the haematite rich bedrock produces a diagnostic magnetic signal for source tracing (Figure 6.5B). Current research suggests that the dominant source of colluvium, alluvium and lake sediments in the Slapton Ley and Start valley sediment sinks is topsoil derived.

Conclusions

In relation to the three questions posed earlier, these preliminary results suggest that the Start valley is acting as a sediment sink causing a major reduction in sediment yield to the lower Start valley and the Ley. However, the rapid accumulation of colluvium behind hedgerows also suggests that fluvial sediment yields may be a significant underestimate of hillslope erosion.

REFERENCES

FOSTER, I.D.L. (1992). *A hydrological and Limnological Survey of the Old Mill Reservoir and Catchment, Dartmouth, South Devon.* Cent. Environ. Res. & Consul. Report, Coventry University.

O'SULLIVAN, P.E., HEATHWAITE, A.L., FARR, K.M and SMITH, J.P. (1989). Southwest England and the Shropshire Cheshire Meres. Guide to Excursion A, Vth International Symposium on Palaeolimnology, Ambleside, Cumbria, UK. 1-6 Sept 1989.

O'SULLIVAN, P.E., HEATHWAITE, A.L., APPLEBY, P.G., BROOKFIELD, D., CRICK, M.W., MOSCROP, C. MULDER, T.B., VERNON, N.J. and WILMHURST, J.M. (1991) Palaeolimnology of Slapton Ley, Devon. *Hydrobiologia* **214**, 115-124.

OWENS, P.N. (1990) *Valley Sedimentation at Slapton, South Devon and its implications for the Estimation of lake sediment based erosion rates.* In: BOARDMAN, J., FOSTER, I.D.L. and DEARING, J.A. (eds) Soil Erosion on Agricultural Land. J.Wiley, Chichester, 193-200.

TROAKE, R.P. & WALLING D.E. (1973) The Natural History of the Slapton Ley Nature Reserve VII The hydrology of the Slapton Wood Stream. *Field Studies* **3**, 719-740.

WALLING, D.E. (1988) Erosion and sediment yield research - some recent perspectives. *Journal of Hydrology* **100**, 113-141.

Chapter 7
Lake sedimentation

Louise Heathwaite
Department of Geography
University of Sheffield

The deteriorating water quality of flowing and standing water bodies has often been linked to land use changes in the drainage basin (Ryding and Forsberg, 1979; Houston and Brooker, 1981; Cameron and Wild, 1984; Ryden *et al.*, 1984; Heathwaite and Burt, 1991). The problem is that from the relatively short (15-20 year) timescales available for water quality records, it is not always possible to discern links between the pattern of land use change and trends in river and lake water quality.

The physical and chemical analysis of lake sediment cores when combined with land use records over longer (50-150 year) timescales can reveal information on the history of a lake and its catchment. At this timescale, it is possible to identify the general trends and driving force in catchment changes as they are passed on to the drainage network and recorded in the lake sediments. Such an approach is not without its own problems, particularly in terms of the indirect link that may exist between a lake and its catchment, and especially where valley sedimentation removes a proportion of the lake sediment source (See Chapter 6). However, in order to determine how water quality deterioration has occurred and over what timescale, lake sediment analysis offers a valuable insight for the palaeoenvironmental reconstruction of lake eutrophication and catchment erosion.

The objectives of the lake sediment research at Slapton have been to match the evidence for accelerated lake sedimentation to land use changes in the catchment and to provide a historical interpretation of lake eutrophication as a result of the increased influx of nutrients from the catchment. A number of different data sources have been exploited in order to reconstruct the past environment of the lake and its catchment. These include chemical fractionation and ^{210}Pb dating of lake sediment cores and 100-year catchment land use and fertiliser records derived from agricultural census returns.

The sediments of Slapton Ley have been previously studied by Crabtree and Round (1967) and Morey (1976). Their results suggest that since its formation by onshore movement of shingle roughly 1000 BP, the Ley has been a shallow, eutrophic lake. Recent evidence suggests that the lake may now be hypertrophic (Heathwaite, 1993; Heathwaite and O'Sullivan, 1991; O'Sullivan and Heathwaite, in press) primarily as a result of increased sediment and solute inputs from its catchment. The current annual load is over 6 t km^{-2} inorganic nitrogen and over 30 t km^{-2} sediment (Heathwaite and Burt, 1991).

Lake morphometry

Slapton Ley is divided into the Higher Ley, a 39 ha reedbed and the 77 ha Lower Ley, which is a shallow freshwater lake with a maximum depth of 2.8 m. The bathymetry of the Lower Ley is shown in Figure 7.1. There are three main inflows to the lake: the Higher Ley at Slapton Bridge, the Start stream and the Stokeley stream. There is one main outflow at Torcross where a sluice artificially controls the water level in the Lower Ley. Seepage also occurs through the shingle ridge. The lake has a flushing rate of approximately twenty times per year, although during high storm flow it is possible that up to one-third of the total lake volume (1.19 x 10^6 m^3) may be displaced in around 24-hours (van Vlymen, 1979).

A field guide to the geomorphology of the Slapton region

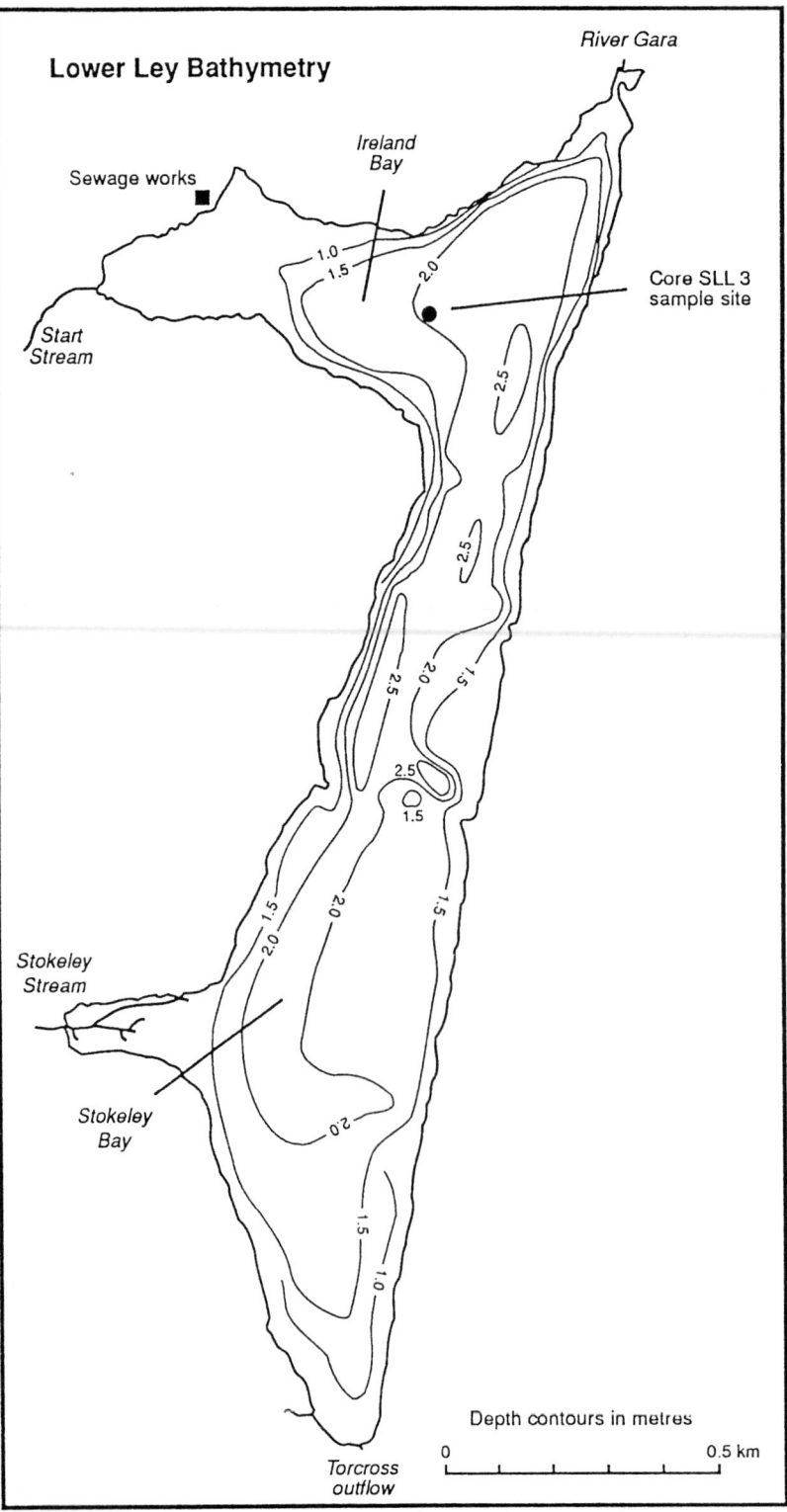

Figure 7.1 Bathymetry of the Lower Ley.

7. *Lake sedimentation*

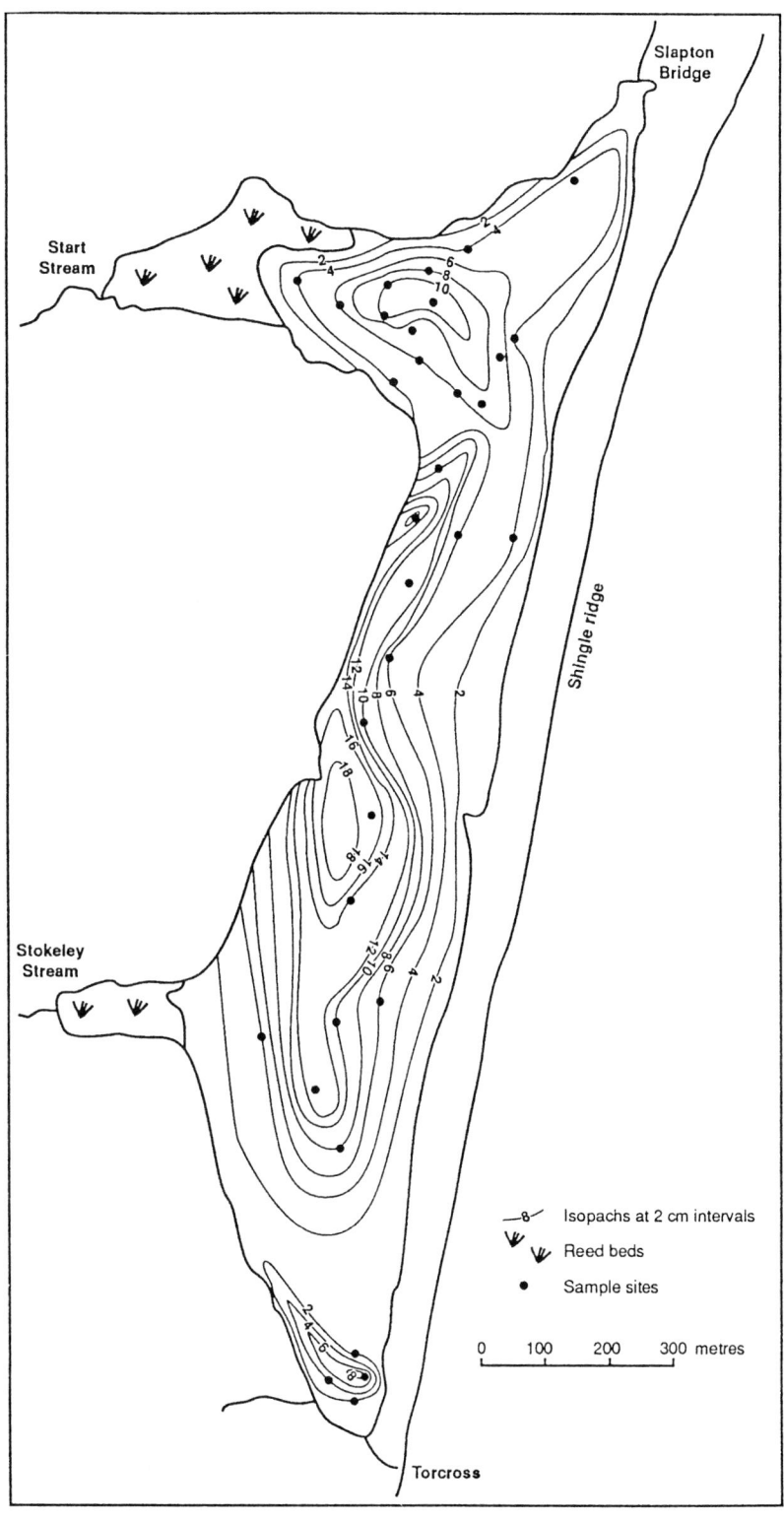

Figure 7.2 The pattern of sedimentation in the Lower Ley. See text and O'Sullivan et al. (1991) for explanation.

The pattern of sedimentation in the Lower Ley

The pattern of sedimentation across the bed of the Lower Ley is shown in Figure 7.2. The data were derived from multiple coring of the Lower Ley combined with the measurement of the whole core magnetic susceptibility of the individual cores (O'Sullivan et al., 1991). This allowed correlation of the pattern of sedimentation across the lake basin for different cores through time and the construction of isopachs of sediment thickness. The pattern of sedimentation is not evenly distributed across the bed of the Lower Ley, and there is some sediment focussing along the long-axis of the lake, away from the main inflows of the Gara and Start subcatchments. Sedimentation rates appear to be lower in the relatively exposed areas of the lake close to the shingle ridge, and higher on the landward site of the Ley.

The pattern of lake sedimentation through time is provided by the ^{210}Pb profile for a core from Ireland Bay (Figure 7.3; the same core was used in the chemical analyses described below). The chronology of the core was derived using the CRS (constant-rate-of-supply) model of Appleby and Oldfield (1978) which assumes a constant net rate of supply of ^{210}Pb from the lake waters to the sediments. This model was used because the unsupported ^{210}Pb profile is of a non-monotonic nature (Heathwaite and O'Sullivan, 1991). The suitability of the CRS model for the Slapton Ley core was confirmed by comparison with the ^{210}Pb profile of another core from the same Ireland Bay location (O'Sullivan et al., 1991). Both cores had comparable total residual unsupported ^{210}Pb-values. The mean sedimentation rate in the Lower Ley is 2.4 mm year^{-1} but there is a significant increase in sedimentation since 1945: the annual rate of sediment influx prior to the Second World War is less than 2 mm year^{-1} increasing to around 10 mm year^{-1} in 1968. At the current sediment-water interface, the rate of sediment influx is around 8-12 mm year^{-1}.

The pattern of sediment influx shown in Figure 7.3 exhibited a strong correlation with the concentration of mineral matter in the core (Heathwaite and O'Sullivan, 1991). This, when taken together with the evidence from multiple coring of the lake, suggests that the amount of eroded material reaching Slapton Ley has increased. On the basis of magnetic determinations, Crick (1985) suggested that the topsoil of local fields, which are rich in haematite, may form the main sediment source. The dry mass content of cores taken from the Lower Ley indicates that on average, 8 t ha^{-1} of dry matter are deposited annually on the bed of the Lower Ley. This is equivalent to a sediment influx rate of approximately 9 mm a^{-1} or an erosion rate from the catchment of over 13 t km^{-2} a^{-1}. This gives a total sediment input in the range 534-734 tonnes over the 77 ha lake area. However, the data from the gauging stations in the catchment suggest that the current suspended sediment load of inflowing streams is around 1440 t a^{-1} (See Chapter 5). This implies that valley sedimentation, particularly in the Higher Ley, is an important sink for sediment eroded from the catchment (Heathwaite and Burt, 1991 - and see Chapter 6).

7. Lake sedimentation

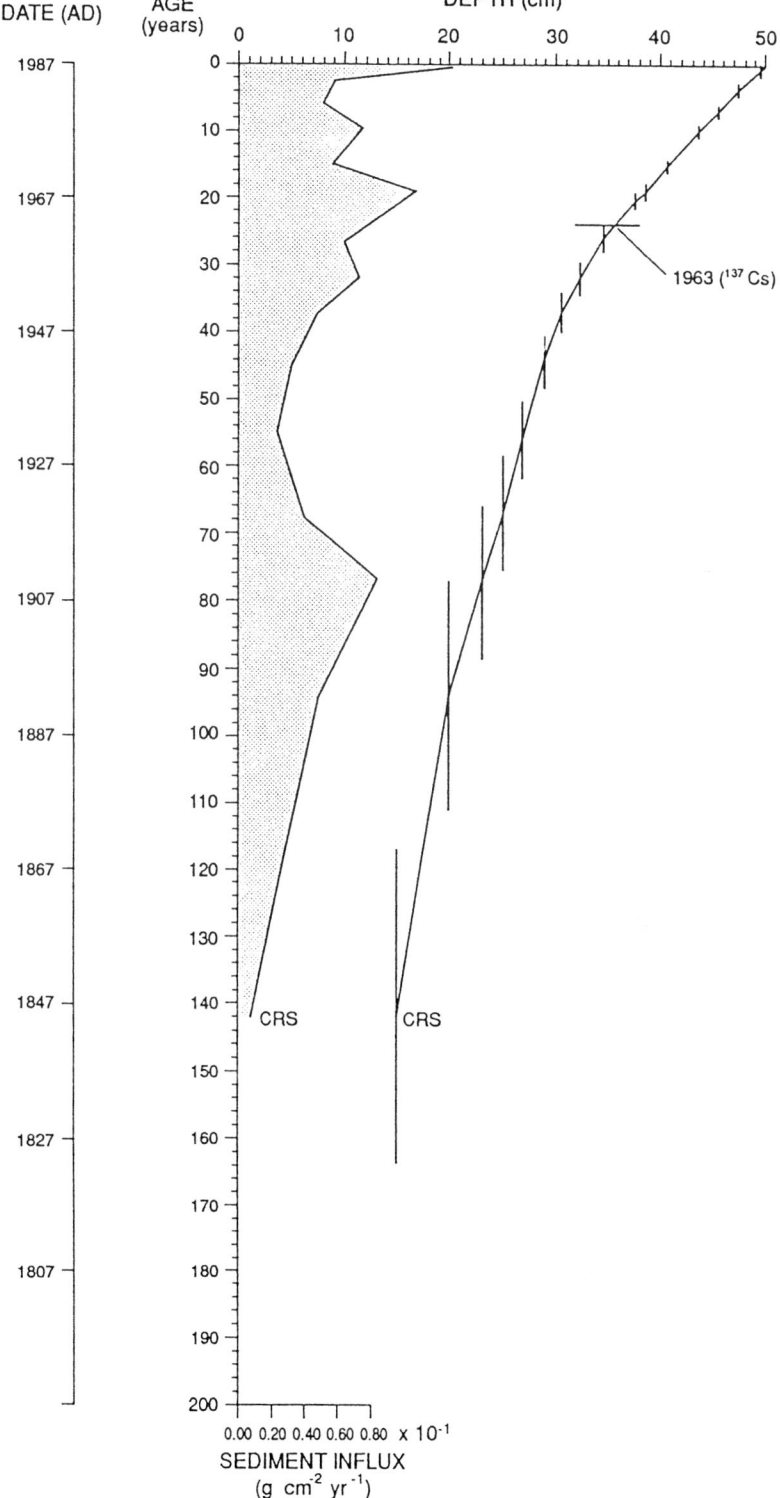

Figure 7.3 Sediment influx to the Lower Ley since c. 1850.

Chemical reconstruction of the history of the lake and its catchment

Most chemical analyses of lake sediment cores have, in the past, focussed on bulk or total element chemistry (Mackereth, 1966). Such an approach gives no information about the source of the sediment material or any alteration in its composition through time, in response to changes in the catchment of the lake. The sediments of the Lower Ley have been analysed using a chemical fractionation technique (after Engstrom and Wright, 1984) to isolate the different sediment fractions into their respective authigenic, biogenic and allogenic origin. Lead (^{210}Pb) dating of the same core (see above) enabled expresssion of the results in terms of sediment influx (annual net deposition of sediment per unit area at the coring site). Because these results are normalised to time, any covariance amongst the different sedimentary components is avoided.

Indicators of catchment erosion

In lake sediments, the allogenic fraction is composed of elements which are embedded in the crystal lattice of the sediment. This fraction therefore represents the influx of largely unaltered and unweathered material to the lake sediments. High allogenic influx is taken as evidence of a high rate of catchment erosion. The elements: K, Si and Al, which are largely associated with silicate minerals, are usually taken as the best indices of catchment erosion (Mackereth, 1966; Engstrom and Wright, 1984). For Slapton Ley, the rate of influx of the allogenic fraction of these elements increases considerably above ca. 21 cm depth, which is dated around 1945 (Figure 7.4). A clear peak in allogenic influx is also shown at 12 cm depth (1965-70). The profiles for these elements suggest that most of the increase in the rate of sediment accumulation since the Second World War is derived from catchment erosion rather than from authigenic material deposited from within the lake water column.

The source of the changes: catchment land use

An explanation for the increase in catchment erosion recorded in the sediment core since 1945 can be sought at two different timescales. First, by examining the long term records for land use change in the catchment and second, using present-day studies of sediment export from key land uses in the catchment (Heathwaite et al., 1990a; 1990b - these results are discussed in Chapter 5). Grassland is the key land use in the catchment, and currently makes up around 85% of the total land area. Between 1955 and 1965, the area of temporary grass increased by almost 1000 ha (or 24% of the total catchment area). This land use shift increased the area of ploughed land in the catchment as land was re-seeded. Thus a greater land area was vulnerable to erosion during this period (Heathwaite and Burt, 1992). Changes in livestock numbers and stocking densities in the catchment may also help explain the increase in the rate of sedimentation recorded in the Lower Ley. Cattle numbers increased from less than 2,000 prior to the Second World War to around 7,000 at present; sheep numbers have increased from 6,000 to almost 25,000. The increase in livestock numbers produced a corresponding rise in the stocking density on permanent grassland from around 5 livestock per hectare prior to the Second World War to nearly 15

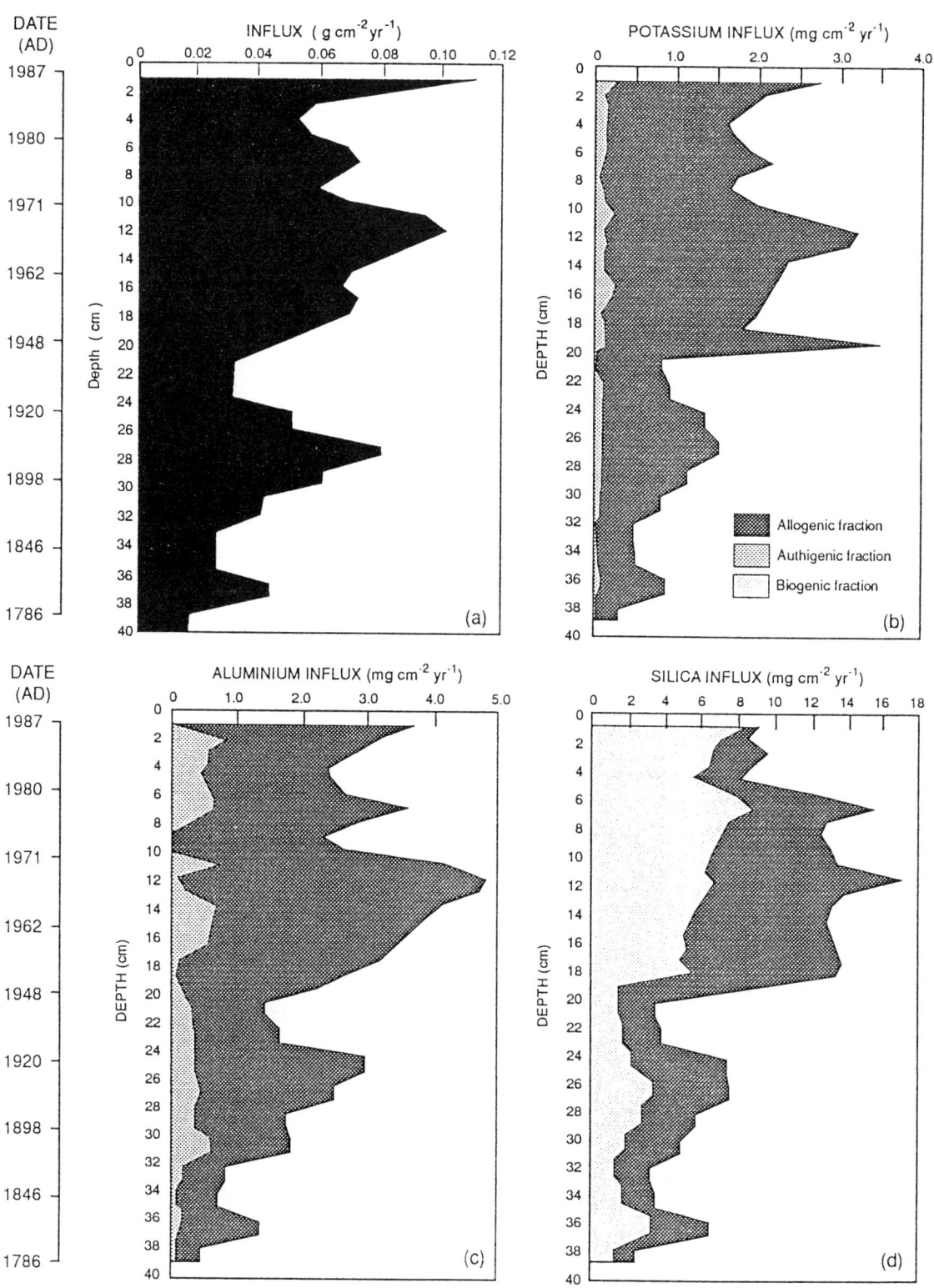

Figure 7.4 Indicators of catchment erosion recorded in lake sediments.

stock per hectare in the early 1950s and again between 1965-1975. Permanent grassland forms at least 60% of riparian zone land use in the Slapton catchment (Heathwaite et al., 1990a). Thus the proximity of this land use, and its associated livestock loading, will increase its relative contribution to the stream sediment and nutrient load (Heathwaite and Burt, 1991; Heathwaite, 1993).

The record of eutrophication recorded in the lake sediments

The increase in the rate of catchment erosion and associated nutrient influx has had a demonstrable effect on the nutrient status of the Lower Ley. Crabtree and Round (1967) found that Slapton Ley has, for most of its existence, been a shallow, clear, eutrophic lake, dominated by macrophytes. Since the 1960s, the lake has been increasingly prone to producing substantial algal blooms (Benson-Evans et al., 1967; Van Vlymen, 1980; O'Sullivan and Heathwaite, in press). Moscrop (1986) recorded a major change in the diatom flora in the upper 6 cm of the lake sediments. The *Fragilaria* which dominated Crabtee and Round's core have been replaced by centric diatoms such as *Cyclostephanos dubius*, *Melosira varians*, *Aulacoseiva granulata*, and also by *Asterionella formosa*. These species are indicative of both more eutrophic conditions (hypertrophy in the case of *C. dubius*) and more turbid, plankton-dominated lake waters. These biotic changes in the lake sediments are reflected in the chemical analyses. Figure 7.5 illustrates the influx of the nutrient elements: nitrogen and phosphorus, to the sediments of the Lower Ley. Also shown is Silica because the biogenic silica fraction is generally taken as an indicator of lake productivity: in the upper 5 cm of the core forms between 90-95% of the total silica influx to the lake sediments. An increase in the influx of authigenic N and both authigenic and allogenic P is recorded for the same period. This reflects an increase in the concentration of these nutrients in the water column of the lake. Dissolved inorganic forms of nitrogen and phosphorus which enter the lake from the catchment are utilised by the lake biota and subsequently deposited to the lake sediments as particulate organic P and organic N which are recorded in the authigenic fraction. Thus the productivity of the lake has increased since the Second World War.

Catchment evidence is available to support the palaeolimnological record for an increase in the rate of eutrophication in Slapton Ley, particularly in the past 30-40 years. For example, the application of inorganic N fertiliser increased ten-fold over the period 1940-1985 (Heathwaite, 1993). At the same time, the organic N and P loading from livestock to grazed grassland also increased. For phosphorus, domestic sewage may be a further source of nutrient enrichment in the lake (Heathwaite and O'Sullivan, 1991).

Conclusions

Sequential chemical analysis of lake sediment cores when combined with land use and water quality records can reveal information on the history of a lake and its catchment. Prior to 1945, Slapton Ley was a shallow, eutrophic but clear water environment. Since the Second World War, agricultural intensification has increased the sediment and nutrient export from the land to the stream. These changes are reflected in the sediments of the lake. Around 1945, there is a clear shift in the

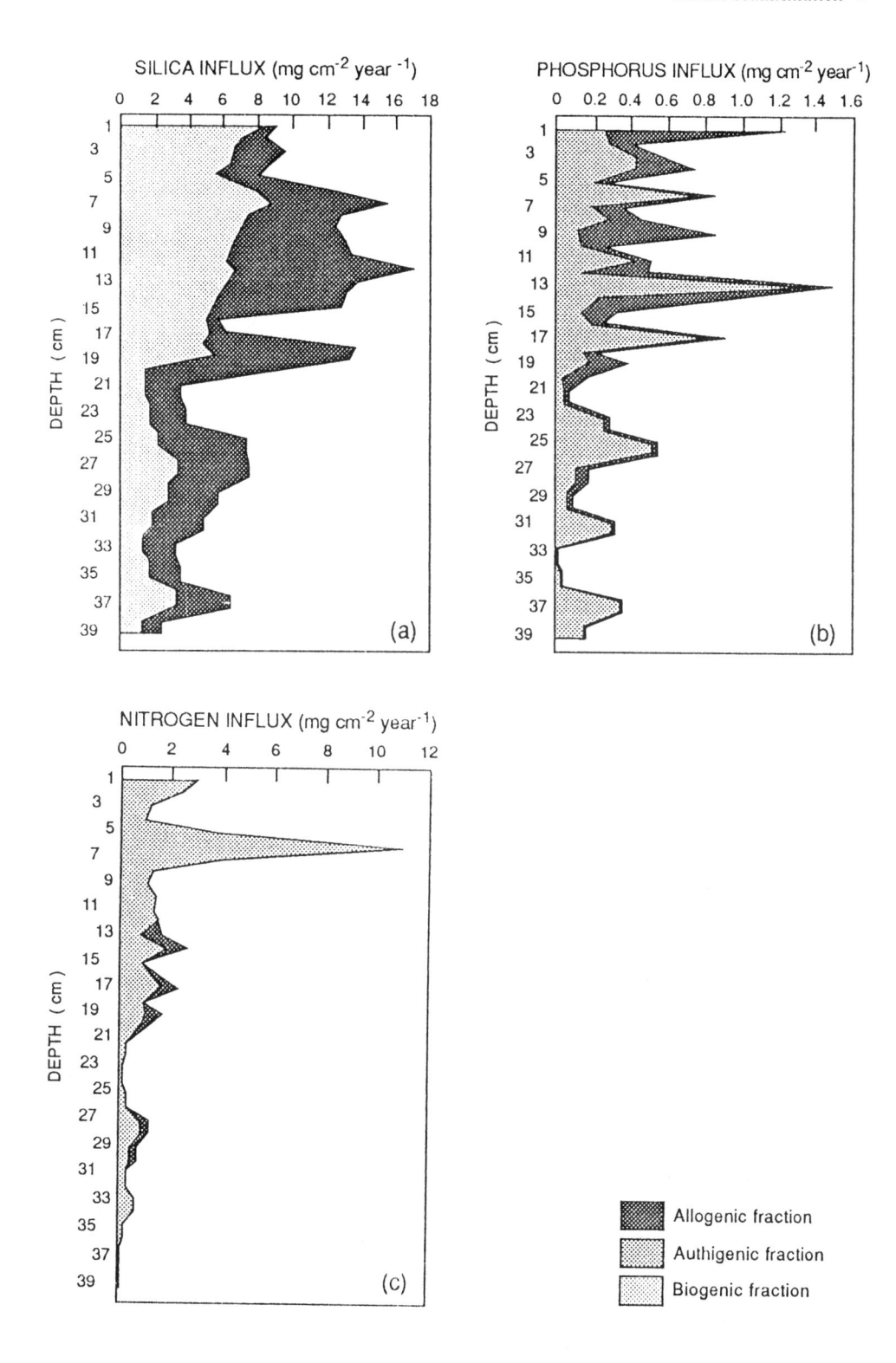

Figure 7.5 *Evidence of eutrophication recorded in lake sediments.*

composition of the lake sediments, and from this time to the present day, the influx of allogenic material from the catchment increased. The source of this material is likely to be sediment eroded from the catchment as changes in the pattern of land use increased the area of ploughed land and the stocking density on permanent grassland. The increased sediment and adsorbed nutrient, particularly phosphorus, loading is recorded in the sediment core as an increase in the proportion of allogenic phosphorus entering the lake.

The land use changes recorded in the agricultural census returns for the catchment show that both the quantity of inorganic fertilisers applied to all land uses in the catchment, and the organic nutrient load voided by livestock, have increased significantly. The increased use of fertilizers (both organic and inorganic) appears to have altered the trophic status of Slapton Ley towards its current hypertrophic position. The shift in the trophic status of the lake is recorded in the accelerated influx of authigenic N and P to the lake sediments and an increase in the productivity of the lake which is reflected in the biogenic silica fraction of the lake sediments.

REFERENCES

APPLEBY, P.G. and OLDFIELD, F. (1978). The calculation of lead-210 dates assuming a constant rate of supply of unsupported 210Pb to the sediment. *Catena*, **5**, 1-8.

BENSON-EVANS, K., FISH, D., PICKUP, G. and DAVIES, P. (1967) The natural history of Slapton Ley Nature Reserve II: Preliminary studies of the freshwater algae. *Field Studies*, , 493-519.

CAMERON, K.C. and WILD, A. (1984). Potential aquifer pollution from nitrate leaching following ploughing of temporary grass. *Journal of Environmental Quality*, **13**, 274-278.

CRABTREE, K. and ROUND, F. E. (1967). Analysis of a core from Slapton Ley. *New Phytologist*, **66**, 255-270.

CRICK, M. W. (1985). *Investigation into the relationship between sediment accumulation in the Lower Ley and spatial patterns of erosion within its catchment.* Unpublished BSc. dissertation, Plymouth Polytechnic.

ENGSTROM, D.R. and WRIGHT, H.E. (1984). Chemical stratigraphy of lake sediments as a record of environmental change. In: HAWORTH, E.Y. and LUND, J.W.G. (eds) *Lake Sediments and Environmental History*, Leicester University Press, Leicester, 11-69.

HEATHWAITE, A.L. (1993). Catchment controls on the recent sediment history of Slapton Ley, southwest England. In: THOMAS, D.S.G. and ALLINSON, R.J. (eds) *Landscape Sensitivity*, Wiley, Chichester, 241-260.

HEATHWAITE, A. L. and BURT, T. P. (1992). The evidence for past and present erosion in the Slapton catchment, southwest Devon. In: BELL, M. and BOARDMAN, J. (eds) *Past and Present Soil Erosion*, Oxbow Monograph 22, 89-100.

HEATHWAITE, A.L. and BURT, T.P. (1991). Predicting the effect of land use on stream water quality in the UK. In: PETERS, N.E. and WALLING, D.E. (eds) *Sediment and Stream Water Quality in a Changing Environment: trends and explanation*, IAHS, **203**, 209-219.

HEATHWAITE, A.L. and O'SULLIVAN, P.E. (1991). Sequential inorganic chemical analysis of a core from Slapton Ley. *Hydrobiologia*, **214**, 125-135.

HEATHWAITE, A.L., BURT, T.P. and TRUDGILL, S.T. (1990a). The effect of land use on the nitrogen, phosphorus and suspended sediment delivery to streams in a small catchment in southwest England. In: THORNES, J.B. (ed) *Vegetation and Erosion*, Wiley, Chichester, 161-179.

HEATHWAITE, A.L., BURT, T.P. and TRUDGILL, S.T. (1990b) Land use controls on sediment production in a lowland catchment, southwest England. In: BOARDMAN, J., FOSTER, I.D.L. and DEARING, J.A. (eds) *Soil Erosion on Agricultural Land*, Wiley, Chichester, 69-86.

MACKERETH, F.J.H. (1966). Some chemical observations on post-glacial lake sediments. *Proc. Roy. Soc. Lond.* B**250**, 167-213.

MOREY, C. R. (1976) The natural history of Slapton Ley Nature Reserve IX: The morphology and history of the lake basins. *Field Studies*, **4**, 353-368.

MOSCROP, C. (1986). *A diatom profile from recent sediments in Slapton Ley.* Unpublished BSc. dissertation, Plymouth Polytechnic.

O'SULLIVAN, P.E., HEATHWAITE, A.L., APPLEBY, P.G., BROOKFIELD, D., CRICK, M.W., MOSCROP, C. MULDER, T.B., VERNON, N.J. and WILMHURST, J.M. (1991). Palaeolimnology of Slapton Ley. *Hydrobiologia*, 214, 115-124.

O'SULLIVAN, P.E. and HEATHWAITE, A.L. (in press) The natural history of Slapton Ley Nature Reserve: The palaeolimnology of the uppermost sediments of the Lower Ley, with interpretations based on ^{210}Pb dating, and the historical record. *Field Studies*.

RYDING, S-O. and FORSBERG, C.(1979). Nitrogen, phosphorus and organic matter in running waters. Studies from six drainage basins. *Vatten*, 35, 46-55.

RYDEN, J.C., BALL, P.R. and GARWOOD, E.A. (1984) Nitrate leaching from grassland. *Nature*, 311, 51-53.

VAN VLYMEN, C.D. (1979). The natural history of Slapton Ley Nature Reserve XIII: The water balance of Slapton Ley. *Field Studies*, 5, 59-84.

VAN VLYMEN, C. D. (1980). *The water balance, physico-chemical environment, and phytoplankton studies of Slapton Ley*. Unpublished PhD thesis, University of Exeter.

Chapter 8
Landforms and Quaternary deposits of the Prawle coast

Derek Mottershead
Edgehill College of Further Education

The Prawle coast possesses a range of geomorphologically interesting features. The rocky shores and cliffs formed by the greenschist exhibit a range of denudational landforms associated with marine erosion and coastal weathering. Resting on the shore platforms and forming the coastal slopes a mass of Quaternary sediments has accumulated, to provide abundant evidence of environmental change.

Bedrock terrain

Shore platforms

Three discrete shore platforms were mapped by Orme (1961) who designated them as 24' OD (7.5m), 14' OD (4.5m) and post Flandrian respectively. Given the slope and consequent range in height of each of these features, and uncertainty regarding their age, they are better termed high, middle and low.

The distribution of the platforms and the relations between them are best observed at low tide when they are well exposed. In general the higher platforms are exposed on rocky promontories, and the lower ones in the bays. Junctions between them are in places marked by an abrasion notch and overlooking cliff. Good examples are present at Sharpers Cove (SX 785358) and Copstone Cove (SX 774381).

The width of the low platform is evident at Gorah (SX 788360) where it was surveyed as 250 m, and viewed through calm water from the cliff above it appears as wide again. This platform is also present at South Hallsands where it has become exposed due to the removal of Flandrian beach sediment since 1903 (Mottershead 1986). Clearly a platform of width up to 500 m is an ancient feature, and even at the rapid rate of formation of 1 cm a^{-1} would require 50000 years to develop. It is therefore likely that this platform is of uncertain Quaternary age, and datable only in relative terms by reference to the overlying sediments.

The weathering of greenschist

The presence of greenschist along the shore and in coastal cliffs creates the opportunity to observe the reaction of this unusual rock in the coastal weathering environment. Freshly weathered rock surfaces are clearly apparent both along the shore in a zone a few metres wide immediately above HWMS and in the exposed coastal cliffs around and to the north of Prawle Point (SX 773349) That contemporary rates of weathering and erosion are rapid is evidenced at several points where human activity has locally served to retard weathering. Three instances can be observed. At Sharpers Cove the remains of an iron stanchion sits embedded in a fillet of cement which stands proud of the greenschist surface, into which it would have originally been set. Secondly, small patches of oil thrown up on the shore have preserved the subjacent rock, whilst the surrounding rock has continued to be lowered by weathering, thus causing oil capped pedestals to emerge from the lowering surface. Thirdly, grease employed in micro erosion meter (MEM) experiments to seal stud sites, has seeped into the adjacent rock binding it and leading to the emergence of visible pedestals in as few as five years.

Active weathering has created a range of distinctive weathering features. In the spray zone the oil capped pedestals noted above attain elevations of up to 11 mm. In planform they are congruent with the oil which caps them, with small and tarry oil patches producing quasi circular pedestals, and runny oil creating less regular planforms. Clearly different oil spills are represented by the pedestals described. Pedestals observed in 1979 were classified according to whether the oil completely covered the pedestal, had frayed back at the edges, or had become thinned and patchy, on the premise that this represented a sequence of ageing and weathering. Median values of pedestal height increase through frayed to patchy, supporting the interpretation of increasing height with age. The majority of pedestals described in Mottershead (1981) are no longer visible in 1992. Other pedestals are now visible suggesting that there exists a continuous process of pedestal generation and decay.

That the rate of weathering in this zone is currently very rapid was confirmed by MEM experiments. A seven year run of data embracing up to 18 measurement points yielded a mean surface lowering rate of 0.625 mm a^{-1} (Mottershead 1989). More detailed analysis demonstrated significant temporal and spatial variatons in weathering rate. Although variations between measurement sites were insignificant, at the microscale within site level, highly significant variations were found. Temporal variations in weathering rate exist at both monthly and seasonal levels. Summer weathering is substantially more rapid than winter, with annualised rates of 1.005 mm a^{-1} and 0.446 mm a^{-1} respectively. The disparity between summer and winter rates was particularly great in 1983 and 1984, both years with hot summers which experienced air temperatures in excess of 24°C. This suggests that temperature is a significant controlling factor on weathering rate.

In order to identify those aspects of the weathering environment responsible for rapid rock breakdown, a series of laboratory experiments was instigated (Mottershead 1982 a, b). In a two way classification experiment rock samples were treated to diurnal immersion in either seawater or deionised water, and allowed either to dry or to remain moist. This experiment demonstrated that alternate wetting and drying in seawater led to rock breakdown. A second set of experiments immersed the rock samples in the individual salts present in seawater, at their respective concentrations. The one effective treatment was wetting and drying in a solution of sodium chloride. It was concluded that of the range of potential agencies in the field weathering environment, wetting and drying in sodium chloride was the agency responsible, through the action of halite crystallisation from the evaporating solution.

There exist on the shore and on coastal cliffs features indicative of rapid weathering. On the shore at Sharpers Cove regular arrays of cylindrical pits at a scale of centimetres form honeycomb weathering patterns (tafoni). They demonstrate an antipathetic pattern to iron impregnation along joints, which stand proud of the intervening pits. Elsewhere small tafoni indent vertical surfaces, penetrating the iron impregnated joint surface and enlarging laterally inwards behind it.

On the coastal cliffs around and to the north of Prawle Point, tafoni at a larger scale are widespread up to elevations of 60m. Often they are measurable in metres and form substantial caverns whose freshly weathered surfaces stand out by virtue of their light colour. Commonly secondary tafonisation is present within them. The tafoni frequently penetrate near-vertical former joint surfaces, on which traces of former iron impregnation remain. They open out laterally and vertically to create lateral and overhanging visors. Secondary tafonisation may penetrate upward into the ceiling of the cavern in the form of cylindrical drills penetrating through the foliation of the rock. The ceilings are coated with a layer of fine granular sediment, which is readily dusted off with a finger. The tafoni walls are commonly coated with a lightly crusted ribbed layer of sediment, a few millimetres deep, which attains the consistency of a slurry when moist. The tafoni floors collect a layer of weathered debris consisting largely of silt and sand with a mode in the coarse silt grade.

The features described point to a process of rock weathering by granular disintegration, and detachment of particles by falling from the ceiling or by slurrying down the walls. The tafoni have evidently developed by recession of the walls both laterally and vertically upwards within the shelter of the caverns.

Tafoni commonly develop in saline environments, either coastal (Hollermann 1975) or hot (Conca & Rossman 1975) and polar (Conca & Astor 1987) deserts, where they are shown to be associated with salt weathering. The Prawle tafoni are consistent with this pattern of occurrence, and it is inferred that salt weathering is responsible for their development; indeed it is very difficult to conceive of any other process which could effect the upward recession of the ceilings in this non-soluble rock. The delivery of sea salts by turbulent onshore winds is certainly capable of depositing salts on the cavern interior walls. Studies of chloride concentration demonstrate decreasing concentration inwards which is consistent with inward penetration of salt from the exposed surface. Extrapolation suggests that chloride may penetrate to depths of 0.1-0.2 m. Salt weathering is therefore capable of providing a satisfactory explanation of both the geometry of the caverns, and the nature of the weathering debris, although the precise mechanism by which it occurs must await further investigation.

Quaternary sediments

Introduction

With the exception of Orme (1961) there has been little attempt systematically to record the stratigraphy of the Prawle coast, or to execute any detailed sedimentological studies. Mottershead (1977) offers a brief review. The sediments observable at any given time depend on the current state of the cliffs, in respect of exposure by recent erosion, or burial by cliff falls.

The sediment sequence

Erratic Boulders: Three igneous boulders of non-local origin are present on the shore platform near Malcombe Point (SX 791362). They currently rest partially buried by contemporary beach, in stratigraphic isolation from other sediments, and offer few clues as to their origin. They are part of a suite of coastal erratics around the English Channel coast extending from Cornwall (Stephens & Synge 1966) to Sussex (Kellaway et al., 1975), Brittany and the floor of the English Channel (Harker 1896). On the north coast of the Cornubian peninsula they are incorporated into Quaternary sediment sequences. The most likely explanation is that they are of western provenance and were rafted up the Channel by icebergs before becoming stranded on the shore.

Raised Beach: Beach deposits are observable at many locations directly overlying the shore platforms. It may vary from well sorted yellow quartz sand with iron concretions as at Mattiscombe (SX 803372) and Lanacombe (SX 815370) where it forms beds up to 2 m deep. Elsewhere raised beach is represented by consolidated pebble beds containing quartz and flint pebbles.

At Peartree Point (SX 820306) a stratum of flint and quartz pebbles including shells of *Patella vulgata* is present within the upper part of the Head deposits over a distance of 150 m. It is 25-30 cm deep and lies 15-25 cm below the top of the section. Orme (1961) postulates a high post Flandrian sea level to account for this feature but, if naturally formed, and in the absence of supporting evidence, it seems more likely to be the result of a single storm episode.

Bedded Deposits: This generic term is used here to embrace a range of hitherto little studied deposits at the base of the Main Head. At many points bedded sand, silts and stony layers form a sandwich of strata at or near the base of the stratigraphic sequence. They appear to represent the local correlative of similar deposits described from Normandy by Watson and Watson (1970). The high fines content may be indicative of an origin as former soils which have stripped by slope processes and redeposited in advance of the main gelifluction episode. One deposit of this group at Malcombe Point was observed in 1979 to contain organic clay with charcoal which has yielded a ^{14}C date (SRR-1855).

Main Head: This deposit forms the main volume of the coastal apron at Prawle. It varies in depth from > 30 m close to the bedrock slope, thinning to ca. 2 m at distant points such as Langerstone Point (SX 782353). These sediments are characterised by local provenance, coarseness, lack of sorting, and angularity of clasts, and have been quantitatively characterised by Mottershead (1971). Often an apparently structureless mass, the Main Head sometimes shows evidence of quasi bedding or coarse lamination. The head is indicative of gelifluction on a massive scale on the coastal slopes. During the accumulation of the Main Head substantial morphological changes were brought about in the local landscape. The head is currently being dissected by streams and trimmed back by coastal erosion.

Upper Head: A bed of finer sediment, 0.3-1m deep commonly overlies the coarse Head. Often buff in colour, it is cohesive and sometimes shows columnar blocky fracture. It is distinguishable in section from topsoil in colour and texture. It is distinguished from Main Head by its low stone content, and a significantly higher silt content of the total fines. The Upper Head is regarded as the local representative of loessic material widely distributed in Devon (Harrod, Catt & Weir 1973) and Cornwall (Catt & Staines 1982). The Prawle material, however, cannnot be regarded as a pure loess; its character suggests mixing by downslope transfer to its present position, probably by gelifluction.

Sandrock: A substantial deposit of consolidated calcareous sand up to 15 m deep is present on the exposed western coast of the Prawle peninsula at Venericks Cove (SX 762364) and Bullock Cove (SX 759364). It exhibits dune bedding and is believed to contain a molluscan fauna. Ussher (1904) seems uncertain as to whether it pre- or post-dates the Main

Head. Karst-like forms on the sandrock have recently been studied by Morawiecka (1993).

Chronology and Dating: Fixed points in the chronology of the Prawle sequence are provided by geochronometric dates. Charcoal in the bedded deposits at Malcombe yielded radiocarbon dates of 17880 ± 260 for the fine fraction, and 22011 ± 240 for the > 125µ m fraction (SRR 1855 a, b). These dates are comparable to dates in the range 21-26000 from basal gelifluction deposits in Scilly (Scourse, 1991). These dates precede the deposition of the Main Head. The Upper Head may be correlated with similar deposits elsewhere in southern England dated by Wintle (1981) by thermoluminescence in the range 14.5-18.5 ka. No dates are currently available for the raised beach or sandrock deposits. In the present state of knowledge, therefore, it appears that a substantial transformation of the Prawle landscape took place in the period approximately 20,000-15,000 years ago, at the time of the main Devensian glaciation elsewhere in Britain. The raised beach clearly predates these events, and the origin of the rock platforms lies further back in time.

REFERENCES

CATT, J.A. & STAINES, S.J.(1982). Loess in Cornwall. *Proceedings of the Ussher Society*, 5.3, 368-375.

CONCA, J.L. & ASTOR, A.M. (1987). Capillary moisture flow and the origin of cavernous weathering in dolerites on Bull Pass Antarctica. *Journal of Geology*. **15**, 151-154.

CONCA, J.L. & ROSSMAN, G.R. (1985). Core softening in cavernously weathered tonalite. *Journal of Geology*, **93**.1, 59-70.

HARKER. A. (1896). Report on a block of stone trawled in the English Channel, and on two erratics near the Prawle, south Devon. *Transactions of the Devonshire Asso*ciation, 531-532.

HARROD, T.R., CATT, J.A. & WEIR, A.H. (1973). Loess in Devon. *Proceedings of the Ussher Society*, **2.6**, 554-564.

HÖLLERMAN, P., (1975). Formen Kavernösen verwitterung Tafoni auf Teneriffa. *Catena*, 2, 385-410.

KELLAWAY, G.A., REDDING, J.H., SHEPARD-THORN, E.R., & DESTOMBES, J-P. (1975). The Quaternary history of the English Channel. *Philosophical Transactions of the Royal Society, London*, A **279**, 185-218.

MORAWIECKA, I., (1993). Palaeokarst phenomena in the Pleistocene raised beach formations of the southwest peninsula of England: preliminary report. *Kras i Speleologica*, 7, (In press).

MOTTERSHEAD, D.N., (1971). Coastal Head deposits between Start Point and Hope Cove, south Devon. *Field Studies*, 3, 433-453.

MOTTERSHEAD, D.N., (Ed.) (1977). *South West England: Guide to excursions A6 & C6*. INQUA. Geobooks, Norwich. 60pp.

MOTTERSHEAD, D.N., (1981). The persistence of oil pollution on a rocky shore. *Applied Geography*, **1**, 297-304.

MOTTERSHEAD, D.N., (1982 a). Coastal spray weathering of bedrock in the supratideal zone at East Prawle, south Devon. *Field Studies*, **5**, 663-684.

MOTTERSHEAD, D.N., (1982 b). Rapid weathering of greenschist by coastal salt spray, East Prawle, south Devon: a preliminary report. *Proceedings of the Ussher Society*, **5**, 347-353.

MOTTERSHEAD, D.N., (1986). *Classic landforms of the south Devon Coast*. Geographical Association, Sheffield. 48pp.

MOTTERSHEAD, D.N., (1989) Rates and patterns of bedrock denudation by coastal salt spray weathering: a seven year record. *Earth Surface Processes & Landforms*, **14**, 383-398.

ORME, A.R., (1960). The raised beaches and stranlines of south Devon. *Field Studies*,1, 109-130.

SCOURSE, J.D., (1991). Glacial deposits of the Isles of Scilly. In: J. EHLERS, P.C. GIBBARD, & J. ROSE (eds.) *Glacial deposits in Great Britain and Ireland*. Balkema, Rotterdam. 580pp.

STEPHENS, N., & SYNGE, F.M., (1966). Pleistocene shorelines. In: G.H. DURY (ed.) *Essays in geomorphology*. Heinemann, London. pp 1-51.

WATSON, E., & WATSON, S., (1970). The coastal periglacial slope deposits of the Cotentin peninsula. *Transactions of the Institute of British Geographers*, **49**, 125-144.

Chapter 9
The Start Bay barrier beach system

David Job
Geography Section, Institute of Education
University of London

The landscape of the Start Bay coast

Looking southwards from the cliff tops above Pilchard Cove (SX837463), the 9km sweep of shingle extending almost to Start Point forms a unique feature of Devon's coastal scenery. On more detailed examination, two distinct environments are evident, one being the detached (or barrier) beaches of Slapton Sands, Beesands and Hallsands (Figure 9.1) where shingle storm beaches impound freshwater lagoons (or leys) on their landward side, the other being the intervening headlands of Limpet Rocks and Tinsey Head and the cliffs at South Hallsands and Pilchard Cove where variable thicknesses of shingle lie at the foot of steep slate or schist cliffs, largely covering the wave cut platforms.

The overall impression is of a series of rias whose entrances have been obstructed by the narrow barrier beaches, impeding the flow of freshwater and sediment to the sea, while the intervening spurs between the rias have been truncated and steepened by marine erosion.

Start Bay itself has an asymmetric shape, described as a zeta-curved bay (Hails, 1975), covering an area of 60km^2. Its floor is a relatively shallow sloping shelf bounded at its eastern margin by a wide submarine bank, known as the Skerries, which runs NNE from Start Point but is separated from the point by a channel of deeper water (Figure 9.1). At low tide, the water depth over the Skerries is sufficiently shallow to cause shoaling of larger waves, while also exerting considerable influence on wave refraction within the bay. To the east of the Skerries, the sea bed drops away more steeply to the deeper water of the Western Approaches.

Sedimentary deposits

Three main sedimentary deposits have been distinguished within the Bay (Hails 1975; Kelland 1975; Morey 1980):

1. Barrier beach deposits. These consist of shingle dominated by rounded flint and quartz gravel (85%) together with smaller amounts of mica-schist, slate and igneous dyke material from local cliffs and rhyolite, felsite, granite and quartz porphyry from more distant sources. They occur as a narrow zone extending from the backshore of the barrier beaches to about 200m beyond low water.

2. Bay deposits. The floor of Start bay has a covering of medium to fine grained sands and silt, greenish grey in colour and rich in shell fragments. Horizons of buried gravel identical with the modern barrier beaches have been interpreted as relict forms of the present day beaches.

3. Bank deposits. The Skerries bank consists of coarse yellow shelly sands which overly the bay deposits and become finer from SW to NE.

Origins of the Start Bay barrier beaches

A range of explanations for the Start Bay barrier beaches have been proposed by a number of workers from the mid-nineteenth century onwards (Pengelly 1870; Kyle 1885; Hunt 1887; Worth 1904, 1909, 1923; Robinson 1962; Perkins 1972; Hails 1975; Kelland 1975; Morey 1980).

Slapton Sands has often been quoted as an example of a shingle spit which has extended across a former coastal inlet by longshore drift to form a bar enclosing a freshwater lagoon on its landward side. Whilst longshore sediment transport is undoubtedly a significant process on the Start Bay coast, most lines of evidence do not support this hypothesis. The shingle lithologies with their preponderance of flint are clearly not derived to any great extent from erosion of cliffs either to the north or south of the beach systems. While schist and slate fragments are present, they are readily reduced by attrition to fines and lost from the beach system in suspension. Quartz pebbles are the only significant component of the beach material which could be attributed to sources from local cliff erosion.

Other writers have proposed that there is an interchange of sediment between the coastal barrier deposits and the Skerries bank by means of a predominant anticlockwise circulation of tidal currents within the bay. It has been suggested that sediment transported northwards across the Skerries by the flood tide is transferred westwards to the coast in the north of the bay then carried southwards along the Start coast by the ebb tide finally returning material to the Skerries offshore from Hallsands. While this mechanism may be responsible for the superficial veneers of fine sediment which cover the shingle at low tide on occasions, the bulk of the barrier beach material is very much coarser and of different lithological composition to the finer bank deposits making up the Skerries.

Early work by Worth and more recent investigations by the Institute of Oceanographic Sciences in the early 1970s (Hails 1975) point to explanations which link the beach material to offshore sources several kilometres east and south of the present coastline and suggest a landward migration of material to its present position during the Flandrian transgression. The following summary sets out the stages proposed in this explanation:

1. A high interglacial shoreline some 7.0m above OD corresponding approximately to the western shores of the present lagoon system now marked by an assumed relict cliff line along the landward shore of Slapton Ley.

2. A low Devensian sea level giving a coastline several kilometres east and south of the present shoreline. Cores from the floor of Start Bay show deeply incised and subsequently in filled channels extending eastwards from the present river valleys. The lithologies exposed to both marine erosion and fluvial activity would have included Tertiary gravels rich in flint, Permo-triassic sandstones and pebble beds rich in quartzite and probably chalk containing flints thus providing sources for the flint and quartzite components of the present barrier beaches. Hails suggests sources 40 km offshore for much of the flint shingle.

3. Carbon dating of organic material from offshore cores suggests a sea level of -40m OD at 10,000 BP followed by a rapid rise to -5m OD at 5000 BP with a mean rate of shoreline advance of about 1km per 1000 years between 8000 and 5000 BP. It is envisaged that, as the shoreline advanced, flint-rich shingle was heaped up as a series of barrier islands backed by shallow lagoons and separated by tidal inlets. These ephemeral barriers then advanced towards their present position overwhelming the estuarine and lagoonal sediments which now lie beneath the modern bay deposits.

4. The occurrence of freshwater peats, tree remains and lake clays beneath the present shingle barriers indicate a final transgression of the shingle barrier into the eastern shores of the freshwater lagoons

9. The Start Bay barrier beach system

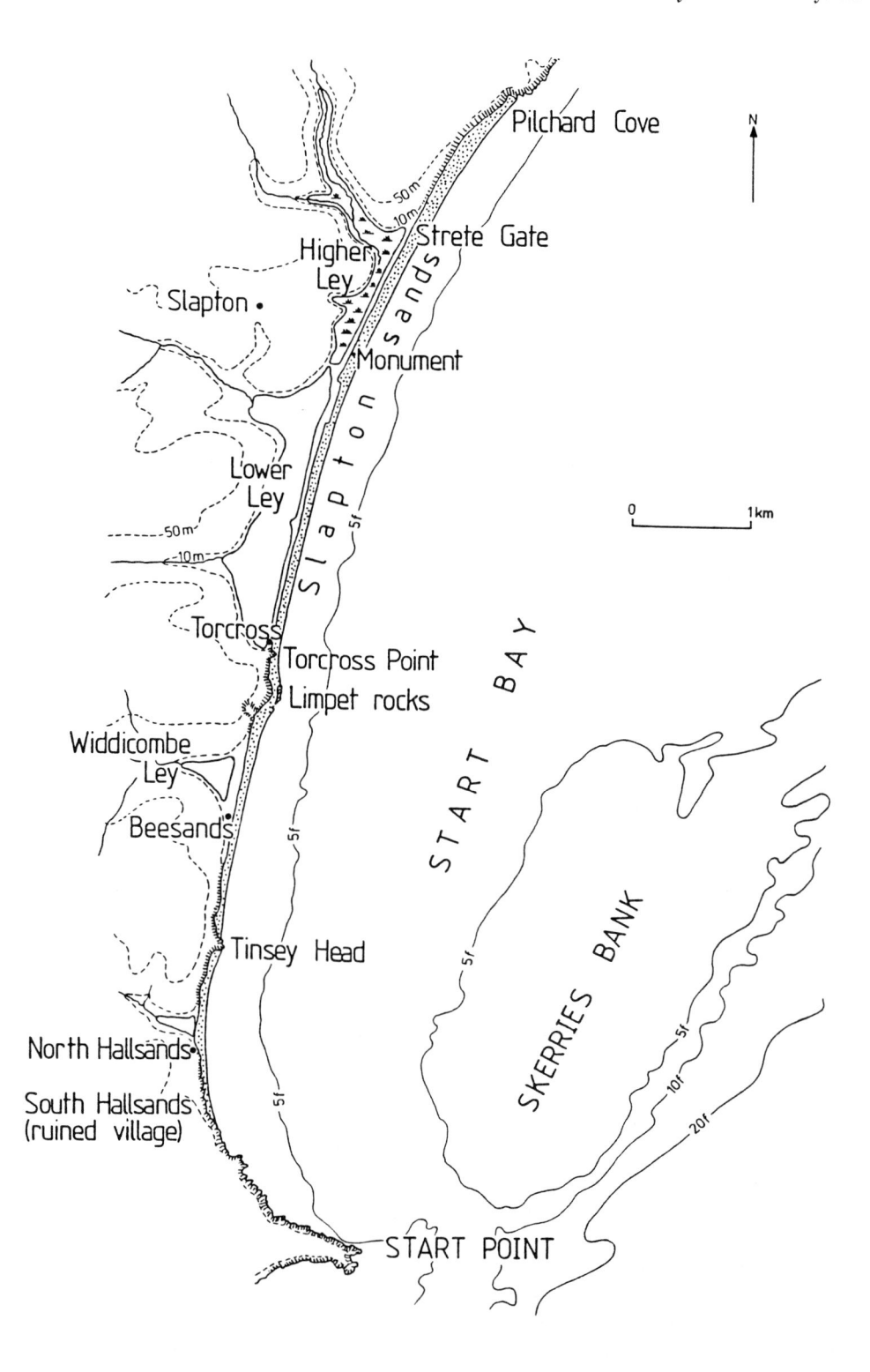

Figure 9.1 Start Bay.

which they enclosed. The oldest dated material beneath Slapton Sands gave a date of 2889 BP suggesting that the barriers were in place by that date.

Kelland proposes that the beach system was already isolated from its main source of supply by about 5000 BP, emphasing the relict nature of the system and the absence of significant natural replenishment under present conditions.

The distribution of shingle along the Start Bay coast

While it can be established that the Start Bay barrier beaches have been in approximately their present position for some 2000 years or more and are therefore in equilibrium with the prevailing long-term wave environment, in the short term their morphology is highly dynamic in response to varying wave conditions.

It is estimated that some 9 million m^3 of shingle fringes the coast between Hallsands in the south and Pilchard Cove in the north (Job 1987). Within this 9km length of coast three barrier beach subsystems can be distinguished linked by shingle-covered foreshores beneath cliffed headlands. For much of the time, longshore transport can occur between the barrier subsystems enabling specific sections of beach to gain or lose sediment according to variations in wave direction. In recent decades a pattern of greater shingle accumulations at more northerly sites has been observed. Beach surveys have been undertaken at intervals throughout the past century at sites along the Start coast (Worth 1904, 1909, 1923 and Robinson 1961 at Hallsands and Gleason *et al.*, 1975 and Job 1987 at sites throughout). Figure 9.2 shows cross sections for sites at intervals along Slapton Sands from Torcross in the south to Strete Gate in the north demonstrating the greater beach width and height northwards. Similar trends are normally evident within the Beesands and Hallsands barriers.

A comparison of monthly surveys at sites along Slapton Sands between 1971/2 and 1981/2 showed a pattern of overall gains at Strete Gate and losses at Torcross and in the central sections of the barrier (Figure 9.3). Photographic and anecdotal evidence indicate a beach width of about 100m at low water at Torcross in the 1930s compared to an average width of about 50m in the 1980s. Tracing experiments have demonstrated that longshore drift occurs from north to south with a NE wave direction and from south to north with a SE wave direction. Clearly with a finite quantity of sediment in what appears to be a reasonably closed system, the two drift directions must be more or less balanced over time in order for the barrier to persist as a continuous feature, but a tendency for northerly drift to dominate in recent decades is apparent. A similar argument could be put forward to account for the drastic decline in beach levels at Hallsands though late nineteenth century shingle extraction is thought to be a more significant factor here with possible repurcussions elsewhere along the Start Bay coast. The depletion of the barriers at their southern ends continues to have important implications for the settlements of Torcross, Beesands and Hallsands, all of which have developed at the southern ends of their respective barriers.

The present pattern of greater shingle accumulations at the northern ends of the barriers may well be a relatively recent phenomenon. Nineteenth century map evidence suggests the reverse to have been the case. Greatest beach width appears to have occurred at the southern ends of the barriers, particularly at Beesands and Hallsands.

9. The Start Bay barrier beach system

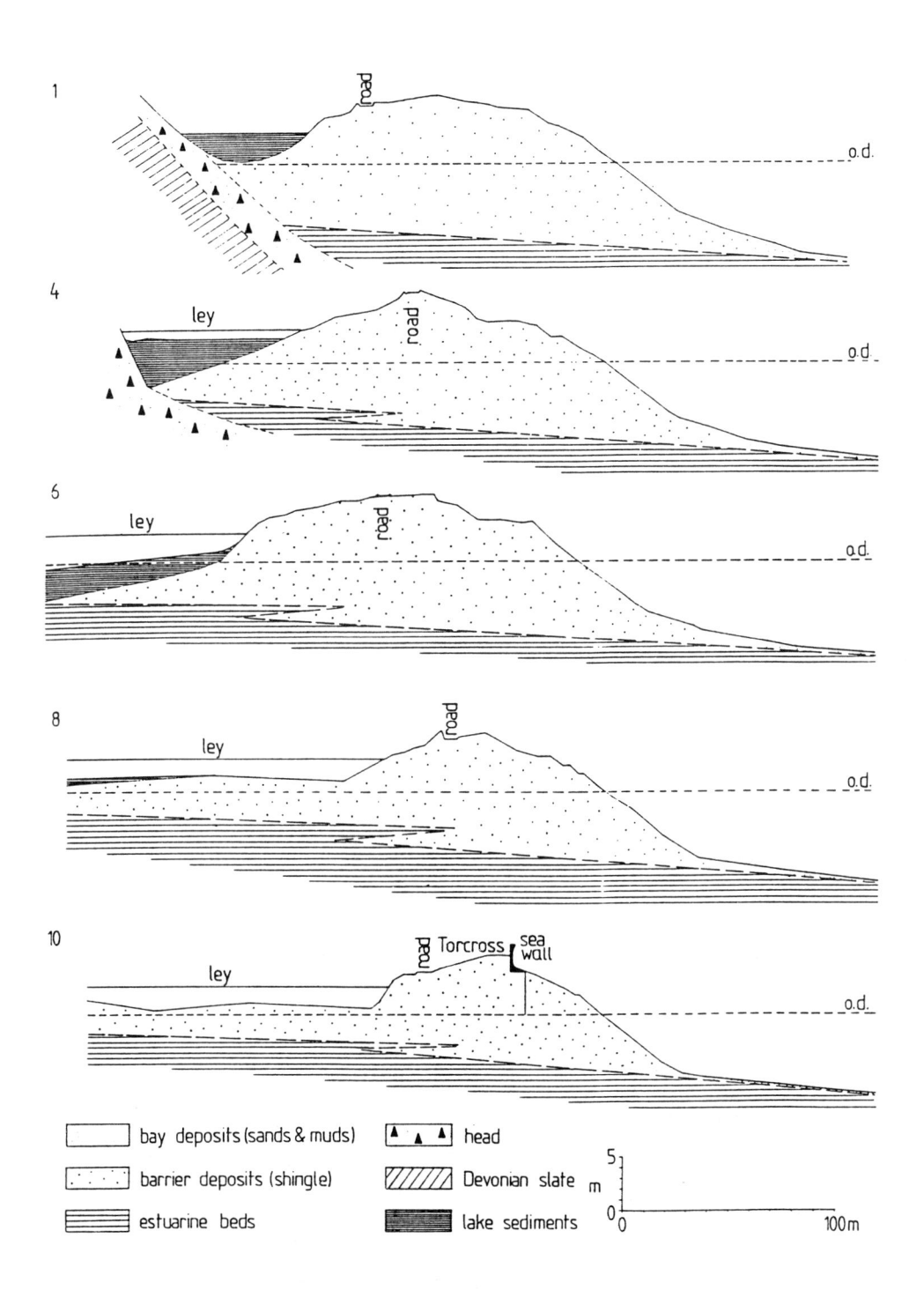

Figure 9.2 Cross-sections across the Slapton Sands barrier beach (site 1 at Strete Gate at the northern end of the barrier, site 10 at Torcross).

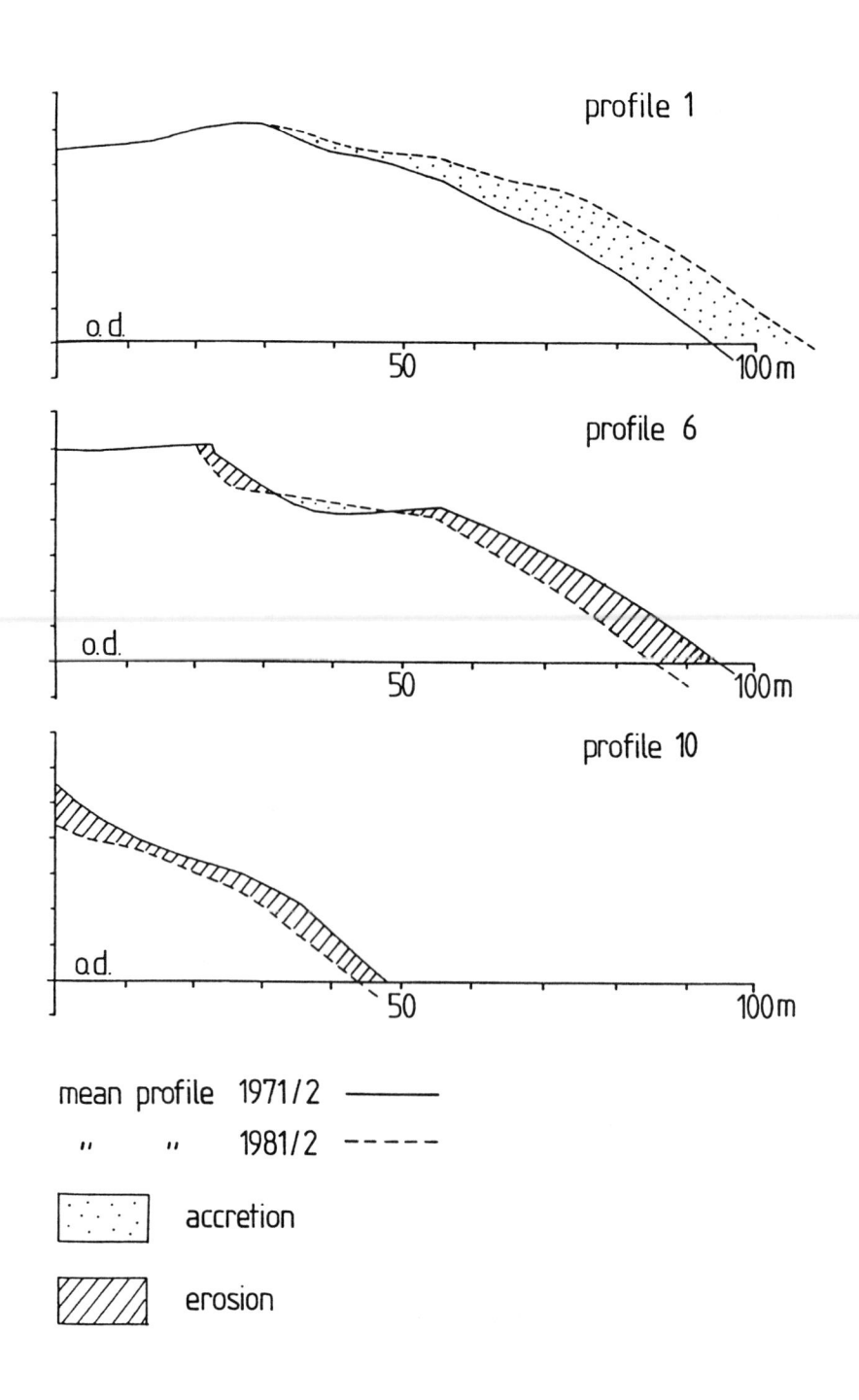

A field guide to the geomorphology of the Slapton region

Figure 9.3 *Redistribution of shingle, 1971/2-1981/2 (profile 1 at Strete Gate, profile 10 at Torcross).*

Short term changes in beach profile morphology

Frequent surveying of beach profiles was undertaken at 10 sites along Slapton Sands between 1981 and 1987 and less frequently at sites at Beesands and Hallsands (Job 1987). Results indicated that much of the variation in shingle levels and profile morphology could not be be related to longshore transport alone in that on many occasions all sites showed either a net gain in beach sediment or a net loss. In most cases changes corresponded well with established models of beach dynamics under varying wave environments. Four distinct wind and wave environments and corresponding profile morphologies were found:-

1. Under conditions of light winds (often offshore), surging breakers of low wave height (<0.3m) generated constructive conditions producing high shingle levels with steep intertidal profiles, a pronounced berm and coarse well sorted surface shingle.

2. Moderate easterly (onshore) winds generated plunging breakers with a higher phase difference (Kemp 1975) which produced net offshore sediment transport, lowering the intertidal beach surface slightly, flattening the profile and destroying the berm.

3. Southerly and south easterly gales generated high (<4.5m), steep plunging breakers of high phase difference oblique to the shore. All profiles were drawn down and flattened with large amounts of sediment transferred to the nearshore zone. Shingle levels at Torcross dropped by up to 3m under these conditions.

4. East to north east gales generated substantial beach accretion on several occasions, a feature contrary to most established models of beach sediment dynamics.

Over the short term, sediment transfers between the beach and the nearshore zone appear to be more significant than longshore transport. The substantial variations in beach level, particularly the spectacular lowering that can occur during southerly gales has important implications for sea defence works. The problems experienced with unsupported boulder rip-rap at Beesands during the 1980s were largely attributable to a lack of appreciation of the magnitude of fluctuating shingle levels.

The coastal settlements of Start Bay

While the more ancient settlements of the Slapton region tend to occupy sheltered inland locations, the three fishing settlements of Hallsands, Beesands and Torcross developed from the sixteenth century onwards in what appear now to be somewhat vulnerable coastal locations. All three settlements seem to have withstood major storms relatively unscathed until the twentieth century, including the so called Great Blizzard of 1891, probably the most extreme storm event on record.

Hallsands The virtual destruction of the old fishing village at South Hallsands (SX 818384) in 1917 has been widely attributed to dredging of shingle from the nearshore zone for the construction of naval dockyards at Plymouth between 1896 and 1901 (Worth 1904, 1909, 1923). The village was constructed on a raised abrasion platform of assumed Interglacial origin (Hails 1975). Prior to 1900 this now seemingly perilous site enjoyed considerable natural protection in the form of a substantial shingle beach which constituted the southernmost extension of the Start Bay barrier. Map evidence, surveys by Worth and a mid 19th century painting by Lidstone all show a wide shingle beach fronting the rock platform on which the village was sited.

Superimposed profiles at the ruined village demonstrate the magnitude of beach losses at Hallsands between 1894 and 1986 (Figure 9.4). The reduced protection afforded by the beach by 1903 accounts for the initial destruction of some buildings including the London Inn as early as 1904. Although stone and concrete sea walls were then built these were overtopped by the 10m waves in the 1917 storms when Worth describes "green walls of water" thundering down on to the thatched cottages. Continued lowering of the shore since may be a response to the scouring action of reflected waves rebounding from the exposed rock surfaces whereas formerly the sloping shingle foreshore dissipated much of the energy of storm waves.

Concern at Hallsands is now focused on the Hallsands Hotel and the adjoining Mildmay cottages (SX 818388) which were constructed on the cliff top to replace the buildings lost in the old village in 1904-5. The cliffs beneath, weakened by shattering and weathering along fault lines normal to the coast, are eroding rapidly in the absence of any significant shingle beach over the rock shore platform beneath. The remaining shingle at Hallsands is now largely confined to a short barrier beach to the north of the Hallsands Hotel which encloses a small ley now colonised by reed and willow.

When SE gales strip shingle from Hallsands beach, there are good exposures of grey clay, reed peat and preserved tree stumps confirming the transgression of the shingle barrier over what were formerly estuarine, lacustrine, reedswamp and finally woodland environments.

Beesands and Torcross The sites of these villages differ from Hallsands in that, while they are located at the southern ends of their respective barrier beaches, they are built largely on the barrier itself rather than a rock foundation. Photographic evidence indicates a far greater beach width at both villages at least until the 1930s. It has been suggested that the depletion of the beaches in recent decades is a longer term response to shingle removal at Hallsands, though the recent tendency for northerly longshore drift to dominate is a possible contributary cause. Both settlements suffered severe damage during storms in January 1979. Villagers campaigned vigorously for sea defences and in 1980 a concrete wave return wall fronted by a rock and concrete revetment with steel piling was constructed at Torcross. A rip-rap defence of limestone boulders was put in place at Beesands but this experienced undermining and collapse when beach shingle was drawn down during southerly gales. This was replaced in 1992 with a concrete wave return wall with steel piling fronted by a broader rock revetment of Scandinavian gneiss.

Conclusion

The Start Bay barrier beaches have been shown to represent a closed system with few present day sources of sediment replenishment. Their formation during the Flandrian transgression from resistant siliceous gravels derived largely from offshore sources is thought to have been complete at least by 2000 BP, isolating the lagoons which they enclosed from direct marine influences. Their dynamic nature in response to changing wave environments and human intervention is clearly shown in dramatic changes over the past century as well as over much shorter timespans. Further work remains to be done in

9. The Start Bay barrier beach system

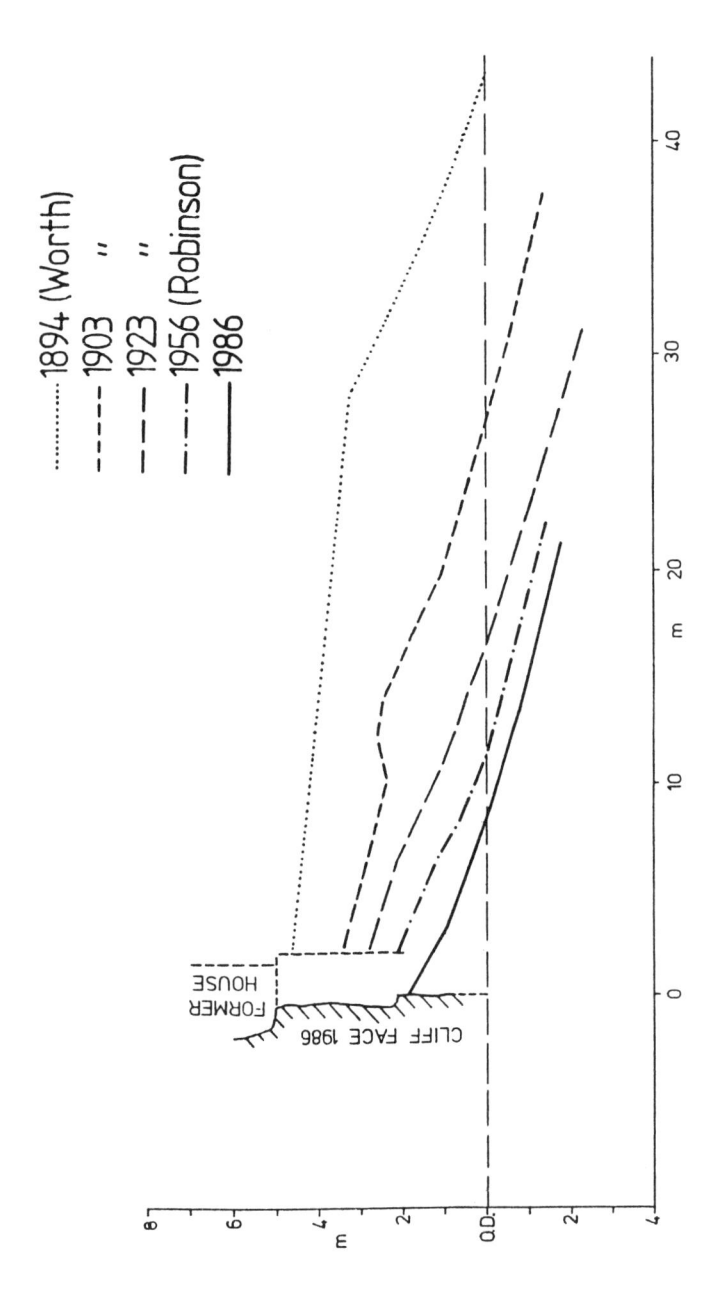

Figure 9.4 Declining beach levels at Hallsands ruined village, 1894-1986.

monitoring future changes particularly to, determine whether recent trends in shingle redistribution are continuing as well as investigating the effects of recent sea defence structures on sediment dynamics. The response of the barriers to any future sea level change would also be of interest and of relevance to the inhabitants of the coastal settlements.

REFERENCES

GLEASON, R., BLACKLEY, M., and CARR, A.P. (1975). Beach stability and particle size distribution, Start Bay. *Journal of the Geological Society, London,* **131**,83-101.

HAILS, J.R. (1975). Sediment distribution and Quaternary history of Start Bay. *Journal of the Geological Society, London,* **131**,19-35.

HUNT, A.R. (1887). The evidence of the Skerries shoal on the wearing of fine sands by waves. *Transactions of the Devon Association,* **19**,498-515.

JOB, D.A. (1987). Beach changes on Slapton Sands shingle ridge. Unpublished M.Phil., Dept. of Geography and Geology, The Polytechnic, Huddersfield.

KELLAND, N.C. (1975). Submarine geology of Start Bay determined by continuous seismic profiling and core sampling. *Journal of the Geological Society, London,* **131**,7-18.

KEMP, P.H. (1975). Wave asymmetry in the nearshore zone and breaker area. in HAILS J. and CARR A. (Eds.) *Nearshore sediment dynamics and sedimentation,* Wiley, London.

KYLE, H.M. (1885). Notes on the physical conditions existing within the line from Start Point to Portland. *Journal of the Marine Biological Association,* **4**.

MOREY, C.R. (1980). The origin and development of a coastal lagoon system, Start Bay, South Devon. Unpublished M. Phil., Dept. of Geography, Portsmouth Polytechnic.

PANGELLY, W. (1870). The modern and ancient beaches of Portland. *Transactions of the Devon Association,* **4**,195-205.

PERKINS, J. (1972). *Geology explained in south and east Devon.* David and Charles, Newton Abbot, Devon.

ROBINSON, A.H.W. (1961). The hydrography of Start Bay and its relationship to beach changes at Hallsands. *Geography Journal,* **127**,63-77.

WORTH, R.H. (1904,1909,1923). Hallsands and Start Bay. *Transactions of the Devon Association,* **36**,302-46; **41**,301-8; **55**,131-47.